Hi-Tech Horticulture

The Author

Dr. Karuna Shanker is currently working as an Assistant Professor and Junior Scientist in Horticulture at Tilka Manjhi Agriculture College, Punsiya, Godda, which is affiliated with Birsa Agricultural University in Kanke, Ranchi, Jharkhand, India. He earned his B.Sc. in Agriculture from Udai Pratap Autonomous College in Varanasi, Uttar Pradesh, followed by an M.Sc. in Horticulture from Acharya Narendra Deva University of Agriculture and Technology in Kumarganj, Ayodhya, Uttar Pradesh. Dr. Shanker completed his Ph.D. in Horticulture from Birsa Agricultural University in Ranchi, Jharkhand, India.

He has qualified the Agricultural Research Service (ARS) through ICAR-NET in Fruit Science and Floriculture, as well as in Landscaping conducted by ASRB in New Delhi. He has authored 18 research papers in national and international journals of repute, 8 book chapters, several popular and semi-scientific articles, and has organized and participated in various training programs, seminars, symposia, and workshops.

Hi-Tech Horticulture

*(AS PER ICAR 5TH DEANS'
COMMITTEE RECOMMENDATION)*

Author

Dr. Karuna Shanker

2025

Daya Publishing House®
A Division of

Astral International Pvt. Ltd.
New Delhi – 110 002

Published by : **Daya Publishing House®**
A Division of
Astral International Pvt. Ltd.
– ISO 9001:2015 Certified Company –
4736/23, Ansari Road, Darya Ganj,
New Delhi-110 002
Ph. 011-43549197, 23278134
E-mail: info@astralint.com
Website: www.astralint.com

Preface

Hi-tech horticulture represents the cutting-edge convergence of technology and agriculture, revolutionizing the cultivation of plants in controlled environments to maximize efficiency, quality, and sustainability. Leveraging advanced techniques such as hydroponics, aeroponics, and vertical farming, hi-tech horticulture transcends traditional farming limitations by optimizing resource use and environmental conditions. Central to this innovation are automated systems that monitor and regulate factors critical to plant growth, including light, temperature, humidity, and nutrient levels. Sensors collect real-time data, enabling precision adjustments that ensure optimal conditions for every stage of plant development. This precision not only accelerates growth cycles but also enhances crop yields, minimizing waste and conserving water and nutrients. Moreover, hi-tech horticulture integrates artificial intelligence and machine learning algorithms to analyze vast datasets and predict crop performance. These technologies empower growers to make data-driven decisions, from crop selection to harvest timing, optimizing productivity and economic returns. Robotics further augment efficiency by automating tasks such as planting, pruning, and harvesting, reducing labor costs and human error.

In urban settings, vertical farms exemplify the potential of hi-tech horticulture to cultivate crops in space-efficient environments, utilizing vertical stacking to maximize land use. These farms often employ LED lighting systems tailored to specific plant needs, creating customized growing conditions regardless of external weather or seasonal changes. Such urban agriculture not only ensures local food production but also reduces transportation costs and carbon footprints associated with traditional farming practices. Beyond economic benefits, hi-tech horticulture promotes sustainability by minimizing pesticide use and soil degradation, fostering healthier ecosystems and safeguarding biodiversity. By enabling year-round production and extending growing seasons, it enhances food security and resilience against climate variability. Hi-tech horticulture exemplifies a transformative approach to agriculture, harnessing technology to address global food challenges while promoting environmental stewardship and sustainable development. As this field continues to evolve, its innovations hold promise for reshaping the future of food production worldwide.

Hi-tech Horticulture have been recommended in the curriculum of B.Sc. (Horticulture) courses of the state agricultural universities. Keeping this in view, the book on has been written for the students. The book has been divided into fourteen chapters covering very comprehensive information on all aspects of Hi-Tech Horticulture including introduction to hi-tech horticulture, hi-tech nursery management and mechanization, micropropagation of horticultural crops, hi-tech field preparation and planting methods, protected cultivation- advantage and constraints, micro irrigation system and its components, canopy management, remote sensing and GIS in horticultural crops, precision farming, precision farming of fruit crops, precision farming of vegetables, precision farming in ornamental crops, mechanized harvesting of produce and post-harvest management of horticultural produce. Related terminology, selected references and web links are also added for the easy understanding and further study on the subject. This book written in simple and understandable language; fully supported by figures and diagrams to help the beginners. This book will be highly useful for students, teachers, professionals, researchers and scholars involved in scientific activities of Hi-Tech Horticulture.

We are grateful to all those persons as well as various books, manuals, periodicals, newsletters, magazines, journals *etc.* that helped in the preparation of this book. In spite of the best efforts, it is possible that some errors may have occurred into the compilation and editing of the book. Further queries, constructive suggestions and criticisms for the further improvement of the book are always welcome and shall be thankfully acknowledged.

Dr. Karuna Shanker

Contents

Chapter 1

Introduction to Hi-Tech Horticulture

Hi-tech Horticulture: An Introduction

Hi-tech horticulture strategies include crop improvement, protected cultivation, mechanization, computerization, post-harvest management, and more. Specific practices covered are integrated pest management, micro-irrigation, plasticulture, greenhouse cultivation, and micropropagation. Horticulture is one of the important components of agriculture sector which has enormous potential to provide food as well as nutritional security. It also can bring prosperity to nation by improving the farmers economic condition and agriculture sector. Thus upgrading the horticultural practices with high technology can fulfil this aspect.

In recent years, the growing population has led to a corresponding rise in the need for food and nutritional security. The traditional farming methods are unable to meet the growing demand, necessitating the implementation of advanced technologies in the agricultural sector. The horticultural sector offers significant opportunities for the application of advanced technology. By growing on small areas of land, particularly for vegetable, flower, and medicinal crops, the yield is higher compared to other types of crops. Hi-tech horticulture has significant potential for achieving high yields and superior quality of agricultural produce. Hi-tech horticulture may be defined as a technology which is increase crop yield by maintenance of environment and soil fertility. They need a lot of money up front, but they help farmers earn more money.

Indian Scenario of Hi-tech Horticulture

Advanced horticulture practices in India Horticulture serves as a way of diversification in the modern day and plays a crucial role in ensuring food and nutritional security, as well as contributing significantly to economic stability. The implementation of horticultural practices has led to progress in several states, including Maharashtra, Karnataka, Kerala, and Andhra Pradesh. Given the significant increase in population, there is a substantial strain on natural resources due to global warming and climate change, diminishing land availability, and a strong demand for high-quality fresh horticulture products. The current situation necessitates a transition to contemporary agricultural techniques, with hi-tech horticulture already at the forefront. Word Hi-tech horticulture encompasses the application of precision manufacturing processes to optimize the utilization of inputs, such as time and quantity, in order to enhance the output and quality of different horticultural crops. It refers to the implementation of advanced technology that is modern, not heavily reliant on the environment, requires significant financial investment, and has the potential to enhance the production and quality of horticulture crops. This is a sequential system of cultivating fruits, flowers, vegetables, and spices, where each step from selecting the seeds/varieties to the final product is carried out using advanced techniques in crop production and post-harvest management. Advanced horticulture techniques have successfully overcome a barrier related to agricultural and climatic conditions. As a result, consumers can now access a wide range of vegetables and other horticultural products throughout the year, albeit at a higher cost. Advanced horticulture techniques encompass several technologies such as Integrated Nutrient Management (INM), Integrated Pest Management (IPM), plasticulture, greenhouse or sheltered growing, hydroponics, microirrigation or drip irrigation, and fertigation.

Hi-Tech Horticulture Technology and Efficient Commercial Production of Horticultural Products

Hi-tech horticulture technology is currently widely used to efficiently produce horticultural products on a commercial scale. They employ numerous advanced horticultural techniques.

- ☆ Integrated Pest Management (IPM)
- ☆ Integrated Nutrient Management (INM)
- ☆ Plastic-culture
- ☆ High Density Planting System
- ☆ Greenhouse Cultivation or Protected Cultivation
- ☆ Hydroponics
- ☆ Micro-irrigation or Drip irrigation
- ☆ Fertigation
- ☆ Sub-surface drainage are examples of hi-tech horticultural practices

Integrated Pest Management (IPM)

In the realm of modern horticulture, where precision, sustainability, and efficiency are paramount, Integrated Pest Management (IPM) stands as a cornerstone practice revolutionizing agricultural landscapes. In the dynamic world of hi-tech horticulture, where cutting-edge technologies intersect with traditional cultivation methods, IPM emerges as a strategic framework aimed at harmonizing productivity with ecological balance. By seamlessly integrating diverse pest management tactics, from biological controls to cultural practices and chemical interventions, IPM offers a comprehensive approach tailored to the unique needs of hi-tech horticultural systems. In this intricate tapestry of innovation and tradition, IPM not only safeguards crop health and yields but also champions environmental stewardship, ensuring a resilient and sustainable future for horticulture at the forefront of agricultural advancement.

IPM may be defined as control the pest by ecosystem approach through different management practices such as crop rotation, mechanical method, physical method, legal control, as well as biological control. IPM is currently a frequently used hi-tech horticulture approach. One of the most important criteria for promoting sustainable agriculture and rural development is integrated pest management in horticulture produce. The goal of integrated pest management is to use a combination of cultural, biological, and chemical methods to control pests and pathogens.

Integrated Nutrient Management (INM)

In the dynamic landscape of hi-tech horticulture, where precision and innovation converge to redefine agricultural practices, Integrated Nutrient Management (INM) emerges as a pivotal strategy in cultivating optimal crop health and maximizing yields sustainably. INM represents a paradigm shift in the approach to nutrient supplementation, transcending conventional methods by integrating a diverse array of inputs, from organic amendments to precision-controlled synthetic fertilizers, tailored to the specific needs of crops and their growing environments. By harmonizing the principles of soil health, plant physiology, and environmental stewardship, INM not only ensures efficient nutrient utilization but also mitigates adverse environmental impacts, safeguarding ecosystems and fostering long-term sustainability. In the realm of hi-tech horticulture, where every aspect of cultivation is fine-tuned for optimal performance, INM stands as a beacon of innovation, guiding practitioners towards a balanced and resilient approach to nutrient management, poised to meet the challenges of tomorrow's agricultural landscape.

In addition, Integrated Nutrient Management (INM) is currently a commonly used Hi-tech horticulture approach. Crop rotation, BNF bacteria (Azatobactor and Rhizobium), PSB, green manure, plant leaf, organic manure, and FYM are some of the strategies utilized to boost crop production without compromising soil fertility. In addition, Integrated Nutrient Management (INM) is currently a commonly used Hi-tech horticulture approach. INM refers to maximizing the advantages from all potential sources of plant nutrients in an integrated way to maintain optimal soil fertility and plant nutrient delivery for sustaining target crop yield. Another key feature of INM is improving fertilizer usage efficiency (FUE) by strategically placing fertilizer in the rhizophere with the most root activity. Today, one of the most frequent approaches among innovative horticulture growers is integrated nutrient management.

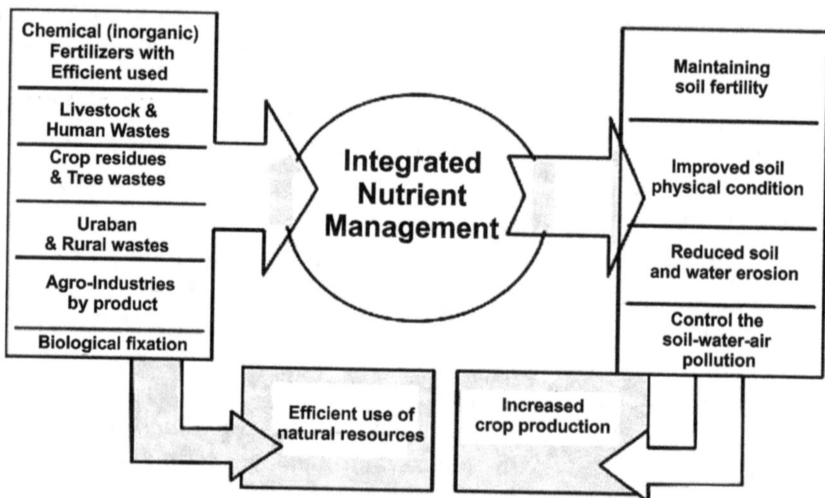

Resources of integrated nutrient management and their role in soil productivity

Plastic Culture

In the realm of hi-tech horticulture, where innovation and sustainability converge to redefine traditional agricultural practices, plastic culture emerges as a transformative methodology revolutionizing crop production. Plastic culture, also known as plasticulture, represents a cutting-edge approach that utilizes various types of plastic materials to optimize growing conditions, enhance crop performance, and mitigate environmental challenges. By leveraging plastics in the form of mulches, tunnels, or containers, horticulturalists can exert precise control over factors such as soil moisture, temperature, and weed suppression, thereby maximizing yields and resource efficiency. In the face of evolving climatic patterns and growing demand for food security, plastic culture stands as a beacon of innovation, offering a versatile and adaptable solution to the complexities of modern agriculture. As hi-tech horticulture continues to push the boundaries of what's possible, plastic culture emerges as a cornerstone practice, empowering growers to cultivate resilient and sustainable crops in an ever-changing world.

Currently, plastic culture is a prevalent advanced gardening approach. Plastics are utilized in diverse ways in the field of commercial horticulture. Plastic culture refers to the utilization of plastics in the context of commercial horticultural output. Plastics are widely used in horticulture for protected cultivation, which includes the usage of greenhouse structures such as high and low tunnels. Applying mulch and encasing with plastic. The implementation of plastic culture improves the economic productivity of industrial procedures and facilitates the efficient utilization of water and energy resources. Plastic culture aids in managing insect and disease infestations while minimizing fluctuations in temperature and moisture levels. Plastic culture significantly contributes to precision irrigation and fertilizer distribution by reducing water and nutrient waste, as well as preventing soil erosion. Plastics have demonstrated their use in promoting the prudent utilization of natural resources, including soil, water, sunlight, and temperature. Utilization of plastic mulch in strawberry cultivation are depicted in Figure 1.1.

Figure 1.1. Utilization of Plastic Mulch in Strawberry Cultivation.

High Density Planting

High-density planting and hi-tech horticulture represent cutting-edge approaches revolutionizing the agricultural landscape. Embracing innovation and precision, these methodologies epitomize the fusion of traditional farming wisdom with modern technology, promising sustainable solutions to the burgeoning challenges faced by the global agricultural sector. High-density planting involves strategically optimizing space utilization by planting crops in closer proximity, thereby maximizing yield per unit area. Conversely, hi-tech horticulture integrates advanced technologies such as precision irrigation, automated monitoring systems, and genetic engineering to enhance crop productivity, quality, and resilience against environmental stresses. Together, these techniques offer a paradigm shift towards resource-efficient, environmentally conscious farming practices, heralding a new era of agricultural productivity and sustainability.

The High Density Planting Technique is a modern method of fruit cultivation that involves planting fruit trees closely together, resulting in smaller or dwarf trees with modified canopy to enhance the absorption and distribution of light, as well as making mechanical field operations easier. The increasing tree density in HDP and meadow orcharding leads to improved production and returns per unit area. Various fruit crops, such as mango and apple, are used for high-density planting (HDP). High density planting in a mango orchard is given in Figure 1.2.

Figure 1.2. High Density Planting in a Mango Orchard.

Protected Cultivation

Protected cultivation, often referred to as controlled environment agriculture (CEA), embodies a sophisticated approach to farming that mitigates the challenges posed by unpredictable weather conditions, pests, and diseases. This innovative method involves growing crops within sheltered structures such as greenhouses, polytunnels, or indoor vertical farms, where environmental factors like temperature, humidity, light, and ventilation can be meticulously regulated and optimized for plant growth. By creating a microclimate tailored to the specific needs of crops, protected cultivation enables year-round production, independent of seasonal variations, thereby ensuring a consistent and reliable food supply. The adoption of protected cultivation offers numerous advantages. Firstly, it extends the growing season, allowing farmers to cultivate crops beyond their natural climatic limitations and thereby increasing overall productivity. Additionally, by shielding plants from adverse weather conditions, such as extreme temperatures, heavy rains, or hailstorms, protected cultivation helps minimize yield losses due to environmental stressors. Moreover, the controlled environment reduces the incidence of pests and diseases, leading to healthier crops and reducing the need for chemical pesticides and fungicides, thus promoting environmentally friendly and sustainable agricultural practices.

Furthermore, protected cultivation facilitates the optimization of resource use, including water, nutrients, and land. Through techniques like drip irrigation, hydroponics, and aeroponics, water usage can be significantly reduced compared to conventional farming methods, while precise nutrient delivery systems ensure optimal plant nutrition. Moreover, the vertical farming approach maximizes land efficiency by utilizing vertical space, making it particularly suitable for urban environments with limited available land for agriculture. In addition to enhancing productivity and resource efficiency, protected cultivation also contributes to food safety and quality. By minimizing exposure to external contaminants and pollutants, such as air and waterborne pathogens, protected environments help maintain higher hygiene standards, resulting in cleaner and safer produce for consumers. Furthermore, the controlled conditions enable growers to manipulate factors such as light intensity, spectrum, and duration to optimize the nutritional content, flavor, and appearance of crops, meeting the increasingly discerning demands of consumers for fresh, nutritious, and visually appealing produce. Protected cultivation represents a game-changing approach to agriculture, offering a sustainable solution to the challenges of food security, environmental sustainability, and economic viability. By harnessing technology to create optimal growing conditions, protected cultivation empowers farmers to overcome the limitations imposed by climate and environmental factors, while simultaneously improving productivity, resource efficiency, and crop quality. As global populations continue to rise and climate change poses ever-greater threats to traditional farming practices, the adoption of protected cultivation is poised to play a pivotal role in ensuring the resilience, sustainability, and security of our food systems in the 21st century and beyond.

- ☆ Green house
- ☆ Net house/shade house
- ☆ Plastic tunnel
- ☆ Rain shelter

Greenhouse Cultivation

Greenhouse cultivation, coupled with hi-tech horticulture practices, represents a formidable alliance in modern agriculture, offering a potent combination of controlled environment precision and technological innovation. Greenhouses provide a sheltered space where environmental variables like temperature, humidity, and light can be finely tuned to optimize plant growth throughout the year. Integrated with hi-tech horticulture, which encompasses advanced techniques such as automated climate control, hydroponics, and sensor-based monitoring systems, greenhouse cultivation achieves unprecedented levels of efficiency and productivity. Through the seamless integration of cutting-edge technologies, greenhouse cultivation and hi-tech horticulture empower growers to produce high-quality crops with minimal environmental impact, ensuring a sustainable and resilient food supply for an ever-growing global population.

Greenhouse or protected horticulture is gaining popularity among progressive horticulturists. This advanced horticulture technology has several advantages compared to traditional production methods, such as greenhouse cultivation. It enables the growth of horticultural products including fruits, vegetables, and flowers in protected surroundings (Figure 3), even when they are out of season. One of the benefits of a greenhouse is the cultivation of vegetable crops.

- ☆ Off-season cultivation of flowers and vegetables.
- ☆ The cultivation of roses, carnations, and other types of cut flowers is being carried out.
- ☆ Plant propagation and seedling rearing
- ☆ Primary and secondary hardening facilities for tissue cultured plants.
- ☆ Rare plants, orchids, herbs, and medicinal plants are cultivated and manufactured.

Figure 1.3. Protected Structure.

Net House/Shade House

Net houses, also known as shade houses, are indispensable components of modern horticulture, offering a controlled environment that strikes a balance between open-field cultivation and fully enclosed greenhouses. These structures consist of a framework covered with specialized shading nets or screens that filter sunlight, providing plants with protection from excessive heat, intense sunlight, and adverse weather conditions while allowing sufficient ventilation. Net houses play a crucial role in mitigating the challenges posed by climate variability, enabling the cultivation of a wide range of crops in regions with harsh environmental conditions. They facilitate optimal growth by regulating temperature and light intensity, promoting higher yields and better-quality produce. Additionally, net houses are cost-effective alternatives to traditional greenhouses, offering growers greater flexibility and adaptability in crop management while reducing energy consumption and environmental impact. In the realm of horticulture, net houses emerge as indispensable tools for sustainable and efficient cultivation practices, empowering farmers to optimize productivity and ensure crop resilience in diverse agricultural landscapes. Net homes or shade houses are utilized for cultivating cucurbits throughout the off-season as they offer protection to delicate plants and supply shade for immature cuttings (Figure 1.4).

Figure 1.4. Net House.

Plastic Tunnel

In the realm of hi-tech horticulture, plastic tunnels stand as iconic symbols of innovation, seamlessly blending advanced technology with traditional farming practices. These structures, also known as polytunnels or high tunnels, represent a sophisticated approach to protected cultivation, harnessing the power of modern materials and design to create optimal growing environments for crops. Consisting of a sturdy framework covered with translucent polyethylene plastic, plastic tunnels offer a controlled microclimate that shields plants from adverse weather conditions while allowing sunlight to penetrate and nurture growth. This fusion of durability, versatility, and affordability makes plastic tunnels indispensable assets in the arsenal of horticulturalists seeking to maximize productivity and crop quality. With the ability to regulate temperature, humidity, and ventilation, plastic tunnels empower growers to extend growing seasons, mitigate climate risks, and cultivate a diverse array of crops with precision and efficiency. As the forefront of agricultural innovation continues to evolve, plastic tunnels remain steadfast pillars of hi-tech horticulture, driving sustainable food production and ushering in a new era of agricultural resilience and prosperity. That is a translucent plastic framework that can be easily repaired. Utilized for the production of vegetables during the non-growing season.

Figure 1.5. Plastic Tunnel.

Rain Shelters

In the landscape of hi-tech horticulture, rain shelters emerge as transformative tools, revolutionizing the way crops are cultivated and protected against the unpredictability of weather patterns. These shelters, meticulously designed and engineered, provide a controlled environment that shields crops from excess rainfall, hail, and other adverse weather conditions, while still allowing beneficial elements like sunlight and air circulation to permeate through. By offering a regulated microclimate, rain shelters enable growers to optimize growing conditions, ensuring that crops receive the ideal balance of moisture and nutrients essential for healthy growth and development.

Constructed from durable materials such as polycarbonate, polyethylene, or even retractable roof systems, rain shelters are engineered to withstand the rigors of outdoor exposure while providing reliable protection for crops. Their versatility allows for customization to suit various crops and environmental conditions, whether in open fields or more controlled settings like greenhouses or polytunnels. Furthermore, advanced features such as automated irrigation systems, humidity control mechanisms, and weather monitoring sensors can be integrated into rain shelters, enhancing precision and efficiency in crop management. The adoption of rain shelters in hi-tech horticulture represents a paradigm shift in agricultural practices, offering growers a means to mitigate risks associated with excessive rainfall, waterlogging, and crop damage. By providing a conducive environment for plant growth and development, rain shelters contribute to higher yields, better crop quality, and increased resilience against environmental stressors. Moreover, they promote sustainable farming practices by reducing water usage, minimizing the need for chemical inputs, and enhancing overall resource efficiency.

As climate change continues to pose challenges to traditional farming methods, the integration of rain shelters into hi-tech horticulture becomes increasingly vital in ensuring food security and agricultural sustainability. These shelters not only protect crops from the vagaries of weather but also empower growers to adapt and thrive in an ever-changing agricultural landscape. In the pursuit of resilient and productive food systems, rain shelters stand as indispensable assets, bridging the gap between nature's unpredictability and human ingenuity in the quest for sustainable food production. A rain shelter is a structure designed to provide protection for crops from rainfall, particularly in hilly regions where rainfall is abundant (Figure 1.6).

Figure 1.6. Greenhouse Practices.

Hydroponics

Hydroponics, a cutting-edge horticulture method, holds significant promise for horticultural producers worldwide. Soilless cultivation is synonymous with hydroponics. Hydroponics enables cultivators to grow plants using nutrient solutions instead of conventional soil. Hydroponics, a revolutionary technique in horticulture, has transformed the landscape of agricultural production by eliminating soil as a medium for plant growth and instead delivering essential nutrients directly to the roots in a water-based solution. This soilless cultivation method offers numerous advantages over traditional farming practices, including greater control over nutrient delivery, optimized resource use, and the ability to grow crops in diverse environments with limited space.

At the heart of hydroponics lies its ability to provide plants with precisely what they need for optimal growth. By delivering nutrients directly to the root system in a controlled solution, growers can tailor the composition and concentration of nutrients to suit the specific requirements of different plant species and growth stages. This precise control minimizes nutrient wastage and ensures that plants receive all the essential elements necessary for healthy development, resulting in faster growth rates, higher yields, and superior crop quality compared to conventional soil-based cultivation. Furthermore, hydroponic systems are highly efficient in their use of resources such as water and space. Unlike traditional farming methods, which often involve excessive water usage and soil degradation, hydroponics recirculates nutrient solutions, minimizing water consumption and reducing environmental impact. Additionally, hydroponic setups can be designed to maximize space utilization, making them particularly well-suited for urban agriculture or areas with limited arable land. Vertical hydroponic systems, for example, allow growers to stack multiple layers of plants vertically, significantly increasing growing capacity without expanding horizontally.

Another significant benefit of hydroponics is its versatility and adaptability to various environmental conditions. Because plants grown hydroponically are not reliant on soil quality or weather conditions, they can thrive in environments where traditional agriculture may be impractical or challenging. This flexibility opens up opportunities for year-round cultivation, regardless of climate or geographical location, thus enhancing food security and resilience in the face of climate change and other environmental pressures. Moreover, hydroponics offers a more sustainable and environmentally friendly alternative to conventional farming practices. By eliminating the need for chemical fertilizers and pesticides commonly used in soil-based agriculture, hydroponic systems reduce the risk of soil pollution and groundwater contamination, promoting healthier ecosystems and safer food production. Additionally, the controlled environment of hydroponic systems minimizes the spread of pests and diseases, further reducing the need for chemical interventions and enhancing crop safety and quality. Hydroponics represents a groundbreaking innovation in horticulture, offering a sustainable and efficient solution to the challenges of modern agriculture. By harnessing the power of technology to deliver precise nutrient solutions and optimize resource use, hydroponic systems empower growers to produce higher yields of healthier, more resilient crops while minimizing environmental impact. As the global population continues to grow and environmental pressures mount, hydroponics stands poised to play a crucial role in shaping the future of food production, ensuring a more sustainable and secure food supply for generations to come. Hydroponic in Tomato is depicted in Figure 1.7.

Figure 1.7. Hydroponic in Tomato.

Micro-irrigation or Drip Irrigation

Drip irrigation is a prevalent irrigation technique employed globally. Drip irrigation offers several advantages compared to conventional irrigation methods. The benefits encompass efficient utilization of irrigation water, enhanced water usage efficiency by directly supplying water to the plant's root system, and reduced loss of soil moisture through evaporation. In the realm of hi-tech horticulture, micro-irrigation, specifically drip irrigation, emerges as a transformative technology that revolutionizes the way water is delivered to crops, maximizing efficiency, productivity, and sustainability. Drip irrigation systems deliver water directly to the root zone of plants through a network of tubes or pipes fitted with emitters, dispensing precise amounts of water and nutrients exactly where they are needed. This targeted approach minimizes water wastage by reducing evaporation, runoff, and deep percolation, making drip irrigation one of the most water-efficient methods of irrigation available today.

One of the key advantages of drip irrigation is its ability to optimize water use while ensuring that plants receive a consistent and adequate supply of moisture. By delivering water directly to the root zone, drip systems minimize water loss to evaporation and surface runoff, allowing for greater water retention in the soil and reducing the risk of waterlogging and soil erosion. This precise control over water application not only conserves water but also enhances crop health and vigor by preventing over- or under-watering, thereby promoting optimal growth and development. Furthermore, drip irrigation systems are highly adaptable and customizable to suit the specific needs of different crops, soil types, and environmental conditions. Growers can adjust factors such as water flow rate, emitter spacing, and irrigation scheduling to optimize water and nutrient delivery for maximum crop performance. Additionally, drip systems can be integrated with other hi-tech horticultural practices, such as fertigation (the simultaneous application of water and fertilizers), to further enhance nutrient uptake and crop yield. Layout of Sprinkler irrigation system is given in Figure 1.8.

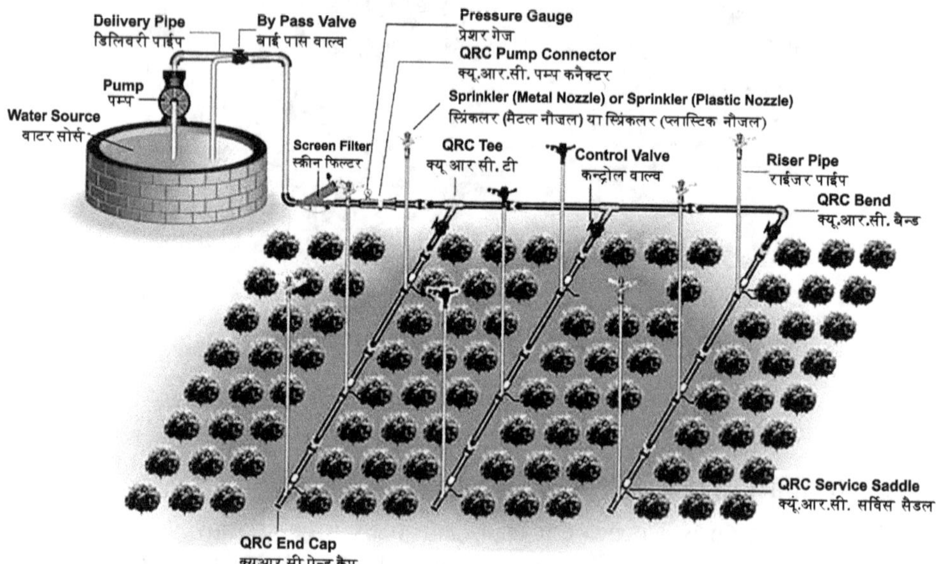

Figure 1.8. Layout of Sprinkler Irrigation System.

Another significant advantage of drip irrigation is its ability to support precision agriculture techniques, such as variable rate irrigation and remote monitoring. By collecting data on soil moisture levels, weather conditions, and crop water requirements, growers can fine-tune irrigation schedules and tailor water application to specific areas within the field, optimizing resource use and minimizing inputs. This level of precision not only improves water efficiency but also reduces energy consumption and environmental impact associated with irrigation practices. Moreover, drip irrigation systems contribute to sustainable agriculture by reducing the need for chemical fertilizers and pesticides. By delivering water and nutrients directly to the root zone, drip systems promote healthier root development and reduce nutrient leaching, thereby minimizing nutrient runoff and groundwater contamination. Additionally, the localized nature of drip irrigation reduces weed growth and pest infestations, leading to lower reliance on herbicides and insecticides, and promoting ecological balance within the agroecosystem. Drip irrigation stands as a cornerstone of hi-tech horticulture, offering a sustainable and efficient solution to the challenges of water scarcity, environmental degradation, and food security. By harnessing the power of technology to deliver water and nutrients precisely where they are needed, drip irrigation systems empower growers to optimize crop production while conserving water resources and minimizing environmental impact. As global populations continue to grow and climate change exacerbates water scarcity and variability, the adoption of drip irrigation is poised to play a pivotal role in ensuring the resilience, sustainability, and productivity of agricultural systems worldwide. Drip irrigation in Banana is given in Figure 1.9.

Figure 1.9. Drip Irrigation in Banana.

Fertigation

Fertigation is the process of delivering plant fertilizers and nutrients by irrigation. Drip irrigation is commonly used for fertilization. In the domain of hi-tech horticulture, fertigation stands out as a groundbreaking technique that merges fertilization with irrigation, offering a precise and efficient method for delivering nutrients directly to plants' root zones. This innovative approach revolutionizes traditional fertilization practices by integrating the application of water-soluble fertilizers into irrigation systems, allowing growers to administer nutrients in a controlled and targeted manner. Fertigation systems typically consist of specialized equipment such as injectors, pumps, and filters, which are integrated into irrigation networks to deliver precise doses of fertilizers along with irrigation water.

One of the primary advantages of fertigation in hi-tech horticulture is its ability to optimize nutrient uptake and utilization by crops. By applying fertilizers directly to the root zone, fertigation ensures that nutrients are available to plants when they need them most, minimizing losses due to leaching, volatilization, or fixation in the soil. This precise control over nutrient delivery allows growers to tailor fertilizer applications to match the specific requirements of different crops, growth stages, and environmental conditions, thereby maximizing nutrient efficiency and crop productivity.

Furthermore, fertigation offers unparalleled flexibility and adaptability, allowing growers to adjust fertilizer formulations and application rates in real-time based on crop responses, soil analyses, and environmental factors. This dynamic approach enables growers to fine-tune nutrient management strategies to optimize crop performance while minimizing nutrient losses and environmental impact. Additionally, fertigation systems can be integrated with other hi-tech horticultural practices, such as drip irrigation and controlled-release fertilizers, to further enhance nutrient delivery efficiency and crop yield.

Another significant advantage of fertigation in hi-tech horticulture is its ability to promote sustainable agricultural practices. By delivering nutrients directly to the root zone, fertigation reduces the risk of nutrient runoff and leaching, minimizing the pollution of water bodies and groundwater resources. Moreover, fertigation allows growers to use water-soluble fertilizers more efficiently, reducing overall fertilizer usage and minimizing the environmental footprint associated with fertilizer production, transport, and application. Additionally, by optimizing nutrient availability and uptake, fertigation helps improve crop nutrient use efficiency, leading to higher yields and better-quality produce with fewer inputs.

Fertigation also enables growers to overcome soil-related constraints and optimize nutrient management in diverse cropping systems, including greenhouse, hydroponic, and field-based production systems. Whether growing fruits, vegetables, ornamentals, or specialty crops, fertigation provides growers with a powerful tool to maximize crop performance while minimizing resource use and environmental impact. Moreover, fertigation systems can be automated and remotely monitored, allowing growers to manage nutrient applications efficiently and effectively, even in large-scale operations or challenging environmental conditions. Fertigation represents a game-changing technology in hi-tech horticulture, offering a sustainable and efficient solution to the challenges of nutrient management, water scarcity, and environmental sustainability. By integrating fertilization with irrigation, fertigation empowers growers to optimize nutrient delivery, enhance crop productivity, and minimize environmental impact, thereby ensuring the resilience, sustainability, and profitability of agricultural systems in an increasingly dynamic and resource-constrained world (Figure 1.10).

Figure 1.10: Fertigation.

Horticulture Business

The main sectors of the economy include horticulture, food processing, fruit and vegetable trade, and floriculture. Hi-tech horticulture and the horticulture business are intricately intertwined, each influencing and shaping the other in a dynamic relationship that drives innovation, productivity, and profitability in the agricultural sector. Hi-tech horticulture refers to the application of advanced technologies, scientific knowledge, and innovative practices to enhance crop production, quality, and efficiency. This can include techniques such as precision farming, controlled environment agriculture, hydroponics, genetic engineering, and digital farming solutions, among others. One of the primary goals of hi-tech horticulture is to maximize crop yields while minimizing resource inputs and environmental impact. By leveraging technologies such as sensors, drones, robotics, and artificial intelligence, hi-tech horticulture enables growers to monitor and manage crops more effectively, optimizing inputs such as water, nutrients, and pesticides to achieve higher yields and better-quality produce. Additionally, hi-tech horticulture allows for more precise and efficient use of land, space, and labor, leading to increased productivity and profitability for horticulture businesses.

Moreover, hi-tech horticulture offers opportunities for diversification and value addition in the horticulture business. By adopting innovative growing techniques and introducing novel crop varieties or specialty products, horticulture businesses can differentiate themselves in the market, capture niche segments, and command premium prices for their produce. For example, greenhouse-grown or hydroponically cultivated vegetables and herbs are often perceived as higher quality and more environmentally sustainable than conventionally grown counterparts, leading to increased demand and profitability. Furthermore, hi-tech horticulture enables horticulture businesses to expand their market reach and distribution channels through e-commerce platforms, direct-to-consumer sales, and value-added products. By leveraging digital marketing strategies and online platforms, horticulture businesses can connect directly with consumers, build brand loyalty, and create new revenue streams. Additionally, hi-tech horticulture allows for the development of value-added products such as processed foods, herbal extracts, cosmetics, and pharmaceuticals derived from horticultural crops, further diversifying business opportunities and enhancing profitability.

Hi-tech horticulture and the horticulture business are mutually reinforcing, with technological advancements driving innovation, efficiency, and profitability in the agricultural sector. By embracing hi-tech horticulture practices and adopting innovative business models, horticulture businesses can enhance their competitiveness, sustainability, and resilience in a rapidly evolving global marketplace. As technology continues to advance and consumer preferences evolve, the integration of hi-tech horticulture and business strategies will play an increasingly pivotal role in shaping the future of the horticulture industry.

Horticulture Food Processing

Currently, the issue of fresh fruits and vegetables spoiling quickly and the subsequent wastage of large quantities of produce continues to be a significant problem. Utilizing horticulture food processing is the one viable solution to reduce food waste. The key to minimizing food waste lies in fruit and vegetable processing. Horticulture food processing constitutes a substantial proportion of the overall food processing industry. Horticultural foods, such as fruits and vegetables, undergo processing to provide a range of value-added goods that can be consumed over an extended period (Figure 1.11). These products include pickles, jams, squashes concentrate, marmalade, fruit blends, canned vegetables, and canned fruits.

Fruit and Vegetable Retailing

Fruit and vegetable retailing is a vital component of the global food industry, serving as the primary link between producers and consumers in the fresh produce supply chain. From local farmers' markets to large supermarket chains, fruit and vegetable retailers play a crucial role in sourcing, distributing, and merchandising a

Figure 1.11. Horticulture Processing Unit.

diverse array of fresh produce to meet the demands of consumers worldwide. One of the key functions of fruit and vegetable retailers is to provide consumers with access to a wide variety of fresh, high-quality produce year-round. Whether sourced locally or imported from distant regions, retailers carefully select and curate their product offerings to ensure freshness, flavour, and nutritional value. This involves working closely with growers, distributors, and wholesalers to procure fresh produce that meets strict quality standards and food safety regulations. Moreover, fruit and vegetable retailers play a pivotal role in promoting healthy eating habits and encouraging consumption of fresh fruits and vegetables. Through creative merchandising, educational campaigns, and nutritional labelling, retailers strive to inform and inspire consumers to make healthier food choices. In recent years, there has been a growing emphasis on sustainability and organic production methods, leading many retailers to expand their offerings of organic and locally grown produce to cater to consumer preferences for healthier, more environmentally friendly options.

In addition to providing fresh produce, fruit and vegetable retailers often offer a range of value-added services to enhance the shopping experience for customers. This may include pre-cut fruits and vegetables, ready-to-eat salads and snacks, and meal kits featuring fresh ingredients, catering to busy lifestyles and convenience-oriented consumers. Many retailers also offer online ordering and home delivery services, allowing customers to conveniently shop for fresh produce from the comfort of their homes. Furthermore, fruit and vegetable retailing is a dynamic and competitive industry, characterized by constant innovation and adaptation to changing consumer trends and preferences. Retailers invest heavily in marketing, branding, and store design to create attractive and engaging shopping environments that appeal to consumers and differentiate their offerings from competitors. This includes strategies such as seasonal promotions, loyalty programs, and partnerships with local growers and community organizations to foster customer loyalty and community engagement.

Overall, fruit and vegetable retailing plays a vital role in ensuring the availability, accessibility, and affordability of fresh produce to consumers worldwide (Figure 1.12). By sourcing, merchandising, and promoting a diverse range of fruits and vegetables, retailers contribute to public health, environmental sustainability, and economic development while meeting the evolving needs and preferences of consumers in an increasingly interconnected global marketplace. The horticultural industry, which involves the sale of fruits and vegetables, is a lucrative sector worth billions of dollars and provides employment to a significant number of small business owners. The retail industry for fruits and vegetables has substantial potential for development in the near future. The consumption of fresh fruits and vegetables is increasing steadily due to heightened consumer awareness of health benefits.

Figure 1.12. Horticulture Product Retailing.

Advantages of Hi-tech Horticulture

Hi-tech horticulture offers a plethora of advantages that contribute to increased productivity, efficiency, and sustainability in agricultural practices. Some of the key advantages include:

☆ **Optimized Resource Use:** Hi-tech horticulture allows for precise management of resources such as water, nutrients, and energy. Technologies like drip irrigation, fertigation, and sensor-based monitoring systems enable growers to deliver the right number of inputs directly to plants, minimizing wastage and maximizing efficiency.

☆ **Increased Crop Yields:** By providing plants with optimal growing conditions and precise nutrient management, hi-tech horticulture can significantly increase crop yields. Controlled environment agriculture (CEA) methods such as hydroponics, aeroponics, and vertical farming enable year-round cultivation and higher crop densities, resulting in greater productivity per unit area.

☆ **Improved Crop Quality:** Hi-tech horticulture techniques enhance the quality and consistency of produce by controlling factors such as temperature, humidity, light, and nutrient levels. This results in fruits and vegetables that are more uniform in size, shape, color, flavor, and nutritional content, meeting consumer preferences and market demands.

☆ **Reduced Environmental Impact:** Hi-tech horticulture practices promote sustainability by minimizing environmental degradation and resource depletion. By optimizing resource use and reducing reliance on chemical inputs, these methods help mitigate soil erosion, water pollution, and greenhouse gas emissions associated with conventional farming practices.

☆ **Year-Round Production:** Controlled environment agriculture techniques allow growers to cultivate crops year-round, regardless of seasonal variations or climatic conditions. This ensures a consistent and reliable food supply, reduces dependence on seasonal imports, and mitigates the risks of crop failures due to adverse weather events.

☆ **Crop Diversity and Specialty Production:** Hi-tech horticulture enables the cultivation of a wide range of crops, including exotic varieties and specialty products that may not be feasible or profitable with traditional farming methods. This promotes agricultural diversity, preserves genetic resources, and supports niche markets and value-added products.

☆ **Enhanced Pest and Disease Management:** Integrated pest management (IPM) strategies combined with hi-tech monitoring systems help growers detect and manage pest and disease outbreaks more effectively. By using biological controls, pheromones, and targeted treatments, growers can minimize reliance on chemical pesticides and reduce the risk of pesticide resistance and environmental harm.

☆ **Labor Efficiency and Automation:** Automation and robotics play a significant role in hi-tech horticulture, reducing labor requirements and improving operational efficiency. Automated systems for planting, harvesting, pruning, and sorting streamline tasks, increase productivity, and reduce labor costs, particularly in labor-intensive crops.

Overall, hi-tech horticulture offers a transformative approach to agriculture, addressing the challenges of food security, environmental sustainability, and economic viability in an increasingly complex and interconnected world. By harnessing technology and innovation, growers can produce more food with fewer resources, while simultaneously improving food quality, environmental stewardship, and economic resilience.

Disadvantages of Hi-Tech Horticulture

While hi-tech horticulture offers numerous advantages, it also presents certain disadvantages and challenges that growers must consider:

☆ **High Initial Investment:** Implementing hi-tech horticulture systems often requires significant upfront capital investment in technology, infrastructure, and equipment. This initial cost barrier may deter small-scale growers or those with limited financial resources from adopting these methods.

☆ **Technical Expertise Required:** Operating hi-tech horticulture systems requires specialized knowledge and technical expertise in areas such as irrigation management, climate control, nutrient optimization, and pest management. Growers may need to invest time and resources in training or hiring skilled personnel to effectively manage these systems.

☆ **Dependency on Technology:** Hi-tech horticulture systems rely heavily on technology and automation, making growers vulnerable to system failures, equipment malfunctions, or disruptions in power supply or internet connectivity. Any technical issues can potentially disrupt crop production and lead to financial losses.

☆ **Energy Consumption:** Controlled environment agriculture (CEA) systems, such as greenhouses and vertical farms, often require significant energy inputs for heating, cooling, lighting, and ventilation. This high energy consumption can contribute to environmental impact and operational costs, particularly in regions with expensive or unreliable energy sources.

☆ **Environmental Concerns:** While hi-tech horticulture can reduce environmental impact in some aspects, such as water and pesticide use, it may also introduce new environmental challenges. For example, the disposal of plastic materials used in hydroponic systems or greenhouse coverings can contribute to waste accumulation and pollution if not managed properly.

☆ **Risk of Overreliance on Technology:** Overreliance on technology in hi-tech horticulture systems may lead to complacency or neglect of traditional farming knowledge and practices. Growers must strike a balance between leveraging technology for efficiency and maintaining a holistic understanding of agronomic principles and ecological dynamics.

☆ **Market Volatility and Competition:** The adoption of hi-tech horticulture methods can lead to increased competition and market saturation, particularly in niche markets or specialty crops. Growers must carefully assess market demand, consumer preferences, and pricing dynamics to ensure profitability and sustainability.

☆ **Social Implications:** The mechanization and automation of agricultural tasks in hi-tech horticulture may lead to displacement of agricultural labour, particularly in rural communities where farming is a primary source of employment. Growers must consider the social implications of adopting technology and explore strategies to mitigate potential negative impacts on local communities.

Overall, while hi-tech horticulture offers significant benefits in terms of productivity, efficiency, and sustainability, growers must carefully weigh these advantages against potential disadvantages and challenges to make informed decisions about adopting these methods. Effective implementation requires careful planning, ongoing monitoring and adaptation, and a holistic approach that considers environmental, social, and economic factors.

Chapter 2

Hi-Tech Nursery Management and Mechanization

Hi-Tech Nursery: An Overview

Managing a hi-tech nursery involves the integration of cutting-edge technology with traditional nursery practices to optimize plant growth, health, and productivity. In today's rapidly advancing agricultural landscape, nurseries equipped with state-of-the-art automation, precision irrigation systems, climate control technologies, and advanced data analytics are at the forefront of innovation. These advancements not only enhance operational efficiency but also enable sustainable practices and ensure the delivery of high-quality plants. By harnessing the power of technology, hi-tech nurseries are revolutionizing the way plants are propagated, nurtured, and distributed, ushering in a new era of precision agriculture.

Nursery is the place where seedling, sapling or any other planting material are raised and sold for planting in the garden or orchard, Whereas, Hi-tech nursery is a place where plants are raised from seeds/other vegetative methods for production of new plants under protected and controlled conditions. All the operations starting from soil preparation to seedling packing in Hi-tech nursery are done with the use of technical knowledge and thus they are expected to deliver good success. Since the propagula get appropriate conditions for growth and development and the practical skill of the grower is assured, hence seedlings and plantlets perform well in Hi-tech nursery. Protected cultivation is intended to mean some level of control over plant microclimate to alleviate one or more of a biotic stresses for optimum plant growth. The microclimatic parameters are temperature, light, air composition and nature of root medium. Success in multiplication under protected conditions increase even in unfavorable agro climatic conditions than open field conditions. Therefore, there is an urgent need for strengthening the concept of hi-tech nursery, where propagation is done under protected condition. Hi tech nursery management includes aspect of micro propagation, Micro-irrigation, fertigation, Hi-tech Greenhouse, Hi-tech plant protection, *etc.*

Greenhouse

In the realm of modern agriculture, hi-tech nursery management and greenhouses represent a transformative approach to sustainable plant cultivation. These integrated systems combine cutting-edge technologies with eco-friendly practices to maximize productivity while minimizing environmental impact. Hi-tech nurseries leverage automation, robotics, and artificial intelligence to precisely monitor and control conditions such as light, temperature, humidity, and nutrient levels. This precision not only optimizes plant growth and health but also conserves resources like water and energy. Simultaneously, greenhouses extend growing seasons and protect plants from adverse weather, pests, and diseases, ensuring consistent yields of high-quality produce. By embracing these innovations, the synergy between hi-tech nurseries and greenhouses not only meets the increasing global demand for food security but also promotes sustainable agriculture practices for a greener future. A greenhouse is a framed or inflated structure covered with a transparent or translucent material in which crops could be grown under the conditions at least partially controlled environment and which is large enough to permit a person to work within it to carry out cultural operations.

Historical Development of Greenhouses

In the 1st century, the cultivation of off-season cucumbers under translucent stones was practiced by the enemy ruler. Tiberius is the earliest documented figure in the practice of protected agriculture. In the 16th century, glass lanterns, bell jars, and hot beds coated with glass were employed to save horticultural crops from the effects of cold weather. Straw mats are commonly utilized in Japan in conjunction with oil paper. A 17th century movable timber frame covered with a translucent paper that has been treated with oil. Heating systems were employed to maintain a warm atmosphere for the plants. Additionally, the usage of glass on only one side, known as a sloping roof, became popular in the 21st century. The use of glass on both sides as a sloping roof began in the 18th century, coinciding with the development of glass houses in England. Protected agriculture was officially founded in 1948 when Professor Emry Myers Emmert at the University of Kentucky introduced polythene as a greenhouse cover. This less expensive material replaced the more costly glass previously employed. Today, in the 21st century, with the advancement of hi-technology greenhouse systems. The production system is highly mechanized, with artificial environmental control and a growing system that is almost completely controlled by computers.

Principle

The yield of a crop is determined by both its genetic makeup and the specific environment in which it grows. The constituents of crop microenvironment are light and temperature. The composition of the air and the characteristics of the root medium. Greenhouse cultivation, commonly referred to as the greenhouse effect, is the ability to manipulate and regulate environmental conditions within a controlled structure.

- ☆ **Light:** Plants rely only on light as their primary source of energy for building tissue and promoting development. While the presence of excessive light itself is not troublesome, the higher temperature caused by intense radiant radiation can result in difficulties associated with high temperatures. In times of difficulty, it is recommended to incorporate shade techniques in the greenhouse. The shade cloth can be operated either manually or automatically through the use of a controller that is regulated by a pyranometer. In winter, several types of lights are used to emit light, such as incandescent, tungsten, halogen, fluorescent, and high intensity discharge (HID) lamps.

- ☆ **Temperature:** Effective temperature control is crucial for ensuring the success of greenhouse crops. Inadequate temperature control can exacerbate disease and result in substandard planting material. Temperature regulation is accomplished by the utilization of diverse technologies such as heating furnaces, exhaust fans, evaporative cooling pads, and shade cloths.

- ☆ **Humidity:** To ensure optimal relative humidity levels in greenhouses, humidification or dehumidification methods are employed. In summer, humidification can be accomplished by using evaporative cooling technologies as fan-pad and fogging systems in combination with greenhouse cooling. During periods of rainfall, both the ambient relative humidity and the humidity inside the greenhouse are elevated. Given the circumstances, the ventilation system is unable to reduce the humidity of the greenhouse air. Therefore, chemical dehumidification systems are employed, which are currently technically viable but come at a high cost.

- ☆ **Water and Nutrients:** Nursery plants cultivated in greenhouses necessitate considerable quantities of water. H_2O is the ubiquitous solvent in plant cells and plays a crucial role in numerous metabolic processes. Water and fertilizer delivery to the plants is facilitated by positioning drip or ring emitters at the plant's base.

- ☆ **Carbon Dioxide:** Studies conducted in northern climes have demonstrated that elevating the concentration of carbon dioxide (CO_2) above the typical ambient range of 350-1000 parts per million (ppm) frequently leads to a higher crop yield. Optimal utilization of this technology necessitates the prolonged closure of houses on a daily basis.

Pest, Disease and Plant Health Management

Effective management of pests and diseases in the greenhouse requires good cleaning practices and timely application of carefully chosen insecticides. Advanced greenhouses are equipped with sophisticated systems to efficiently monitor and detect insect pests or diseases that may affect plants. Appropriate equipment and chemical formulations are then used to manage plant health issues. Biological control techniques have gained more acceptability for plant protection in greenhouses in recent times.

Control Systems

Traditional greenhouses rely on a combination of manual changes, timed events, and theoretical regulations to control the environment. However, modern computerized environmental control systems enable the integration of many greenhouse components, resulting in a more efficient and lucrative system. Automated control systems ensure consistent and optimal environmental conditions. Automating irrigation systems enables labor efficiency and precise control over the timing and quantity of irrigation. Similarly, controllers have offered versatility in controlling heaters, ventilation fans, and wet pads. Computerized control systems can enhance a grower's overall management approach by offering consistent and comprehensive data regarding the greenhouse environment.

Different Protected Structure

There are various types of protected structures, ranging from cloches and frames to hi-tech greenhouses, which can be described as follows.

Cloches

Cloches create a favorable microclimate that can be utilized to advantage plants at the same developmental stage. Plants cultivated beneath cloches will have a faster and more accelerated growth compared to those planted outside without any protection.

Types of Cloches

There exist multiple varieties of cloches, and regardless of their size or shape, it is imperative that they are securely fastened to the ground.

☆ **Glass or Plastic:** Traditionally, cloches were mostly constructed from glass. However, due to the significant risk of breakage, many now choose for the use of stiff, white translucent polypropylene sheeting as a safer alternative. Additionally, thin plastic tubing or sheeting is employed. Glass has the ability to capture and retain long waves reflected by the sun, unlike plastic. As a result, the climate inside glass cloches tends to be warmer, on average, compared to plastic ones.

- ☆ **Tent:** These consist of two sheets of 24-gauge glass measuring approximately 22 x 60 cm2 (9 x 24 inches) that are secured together using specially galvanized wires. This layout provides a narrow width of 30 to 38 cm (12 to 15 inches) at the base, which severely limits the ability to crop.

- ☆ **Barn:** These objects have a shape that resembles a barn. Essentially, there are two glass sheets measuring around 60 x 30 cm2 (24 x 12 inches) on the roof, along with two glass sheets measuring 60 x 15 cm2 (24 x 6 inches) that make up the sides. Once again, wires are employed to fasten the glass together. This design produces cloches that have a width of 60 cm (24 inches) and a height of 23 cm (9 inches) at the center. End sections are often accessible, and certain barn cloches can be equipped with additional side pieces to increase the height to 36 cm (14 inches).

- ☆ **Plastic tunnel:** These are created by elongating transparent plastic sheeting over wire hoops. The hoops are constructed from robust fencing wire with a thick diameter, which is precisely cut into segments measuring 150 cm (5 ft) in length. An "eye" is formed at a distance of 15 to 20 cm (6 to 8 inches) from each end of the hoop, and the hoop is then curved into a semicircular shape. To construct the cloche, position the hoops at a distance of 60 cm (2ft) from each other in a linear arrangement. Place a wooden stick, measuring 38 to 45 cm (15 to 18 inches) in length, at an oblique angle and approximately 40 cm (2 feet) away from the end hoops, at both ends of the row. Securely fasten one end of a 120 cm (4ft) wide sheet of 38 micron transparent plastic to one stake, gently extend the plastic over the hoops, and firmly knot it to the second stake. Securely position the plastic at each hoop by threading thin wire through the openings. When in a windy location, pile earth around the perimeter of the plastic sheeting.

Frames or Miniature Greenhouse

Frames function as small-scale greenhouses, allowing light to enter and capturing solar heat. They are utilized to advance the growth, cultivation, and reproduction of plants through seeds or cuttings. They can be utilized both outdoors and within a greenhouse. These structures are frequently employed for the purpose of propagation. The foundation of a frame is often constructed from wood, masonry, or a metal alloy and exhibits a greater height in the rear compared to the front. The Dutch frame is around 155 × 63 cm (62 x 25 inches) and is constructed with a wooden frame holding a single sheet of 24-gauge glass. The glass may be smoothly moved along the bottom and top rails. The object is fixed in place by a small wooden beam fastened with galvanized nails. There are no specific restrictions for the size of the frame, although a common size is 150 by 80 cm (5 by 2 ½). Frames can be constructed using materials such as brick, composition blocks, wood, or asbestos. Brick and composition block walls offer superior insulation compared to wood and are recommended for frames that require heating for a significant portion of the year. The optimal height for the base of low-growing vegetables, such as lettuce, flower crops, and bedding plants, is 30 cm at the back and 23 cm at the front. The recommended height for taller plants is 45 cm at the back and 38 cm at the front. An optimal frame can typically be achieved by adjusting the light angle to a more vertical position. In order to accomplish this, you will want two lights. One light should be positioned at the optimal angle to capture the sun's rays, while the other light should be placed horizontally at the top of the frame to enhance its overall depth. For most activities, it is customary to have frames oriented towards the south. The frame is particularly intended for shade-loving plants.

Lean to Greenhouse

These types of greenhouses are affixed to the side of a structure and should be positioned on the southern side of the building. In order to optimize the utilization of sunlight and reduce the need for roof supports, the plant is designed to function efficiently. The building's roof is expanded using suitable greenhouse covering material and the space is securely enclosed.

Tunnel Type of Greenhouse

To provide temporary protection against cold and frost, one can construct tunnels approximately 6 feet in height by bending steel tubes or bamboo and covering them with polyethylene sheets. The polythene cover can be removed when the weather conditions are conducive.

Shade Net House

These greenhouses are designed to reduce solar radiation and protect leaves from blistering and withering due to high temperatures and intense sunlight.

Quonset Type Greenhouse

This sort of greenhouse will feature a semi-circular roof. Polythene, fiberglass, or polycarbonate sheets can be used to cover the roof and sides. The optimal dimensions for the greenhouse are a length of 96-120 feet, a width of 29 feet, side walls of 8 feet in height, and a peak height of 12 to 14 feet. The greenhouse is separate from other structures and has a basic framework. The frame primarily comprises of a curved pipe that forms the truss and Quonset outline of the greenhouse.

Even Span Greenhouse

These kinds of greenhouses are built in flat parts of India. This compact greenhouse is specifically designed for use on level ground. This greenhouse is built with two roof slopes that have the same pitch and width. These can exist in several forms, either as single spans or multiple spans. The typical length for a single span generally ranges from 5 to 9 meters, with an average length of approximately 24 meters. The height ranges from 2.5 to 4.3 meters.

Uneven Span Greenhouse

This greenhouse is built on uneven topography. The greenhouse roofs have varying widths, allowing the owner to utilize the inclined sides of the hill. They have a high level of adaptability in mountainous terrains. These cannot be easily automated.

Figure 2.1. Different Protected Structure.

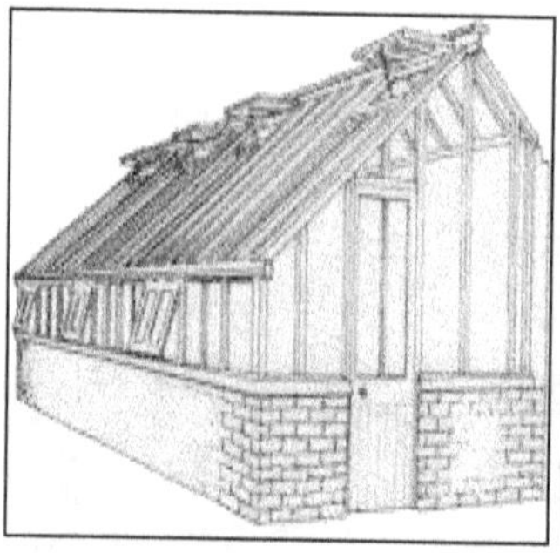

Uneven Span Type

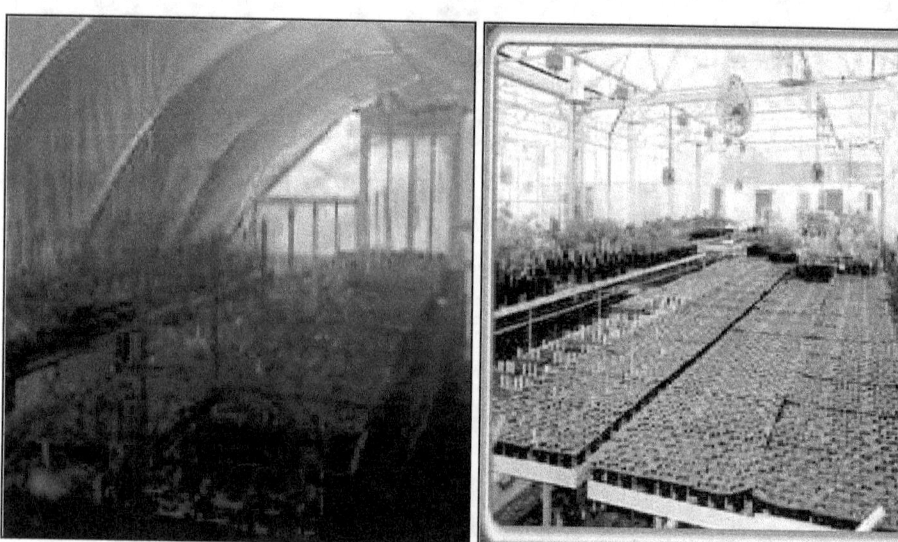

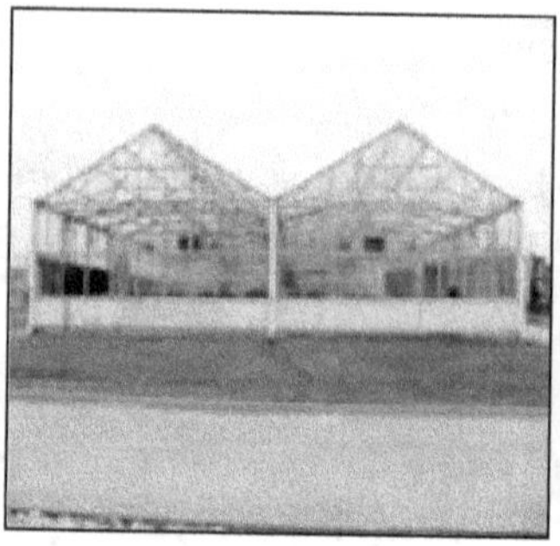

Ridges and Furrows

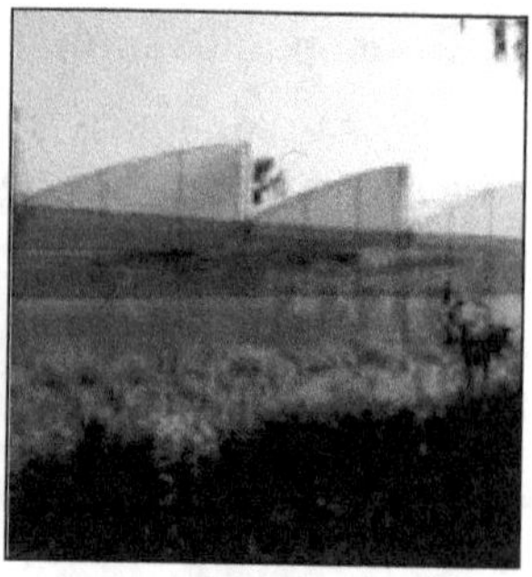

Saw Tooth Type

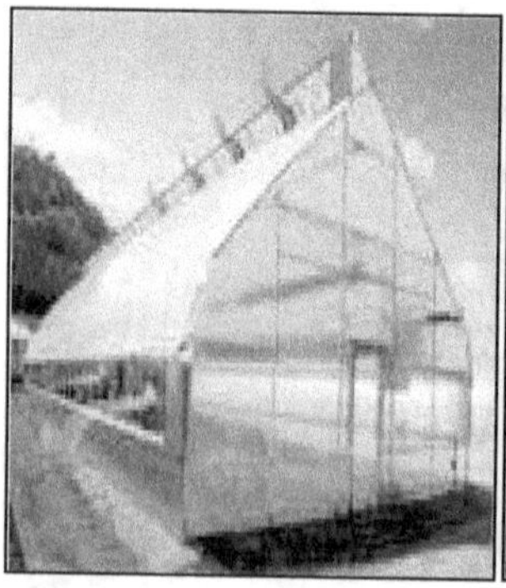

Naturally Ventilated

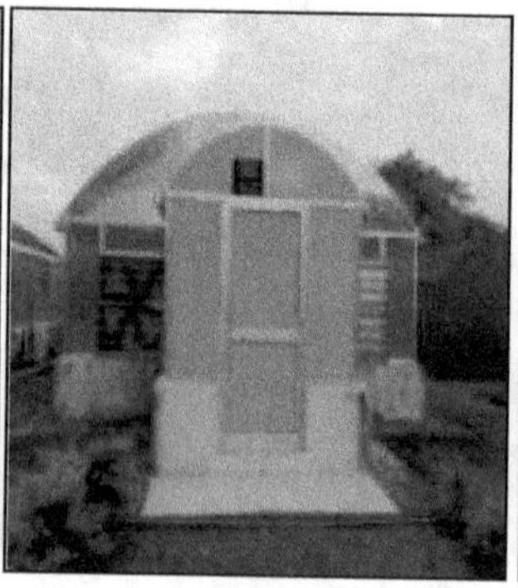

Fan Type

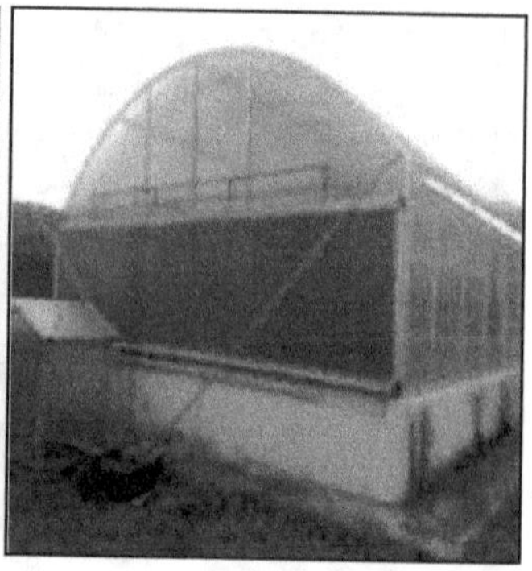

Pad Type

Mist House

Hi-tech Greenhouse Different

Figure 2.2. Different Protected Structure.

Ridges and the Furrow Type Greenhouse

This greenhouse is built on uneven topography. The greenhouse roofs have varying widths, allowing the owner to utilize the inclined sides of the hill. They have a high level of adaptability in mountainous terrains. These cannot be easily automated.

Saw Toothed Type Greenhouse

These structures are comparable to ridge and furrow greenhouses; except they are designed to allow for natural ventilation. A distinct airflow pattern is created in a greenhouse with a sawtooth roof design, allowing for natural ventilation.

Naturally Ventilated Greenhouses

This particular type of greenhouses lacks any environmental control systems, with the exception of sufficient ventilation. These greenhouses serve the aim of safeguarding plant material from adverse weather conditions such as rain, frost, hail, as well as protecting against insects and illnesses.

Partially Environmental Controlled Greenhouse

There is limited control over the microclimate of the plant. Multiple microclimatic components in these greenhouses are regulated utilizing various control systems. Some examples of this style of greenhouse include the Mist house and the Fan and pad type of greenhouse.

Hi-tech Greenhouse

These greenhouses incorporate a control system, sensors, computer control, and an operator. As the name suggests, these greenhouses utilize advanced technologies and systems. Depending on the level of sophistication, the environmental control system in a greenhouse may have partial or complete control over microclimatic parameters. To effectively manage the environmental control equipment.

Table 2.1. A General List of Equipment for Greenhouse Environment Control

Component	Parameter Controlled/Modified
Lighting system	Supplemental light, photoperiod and temperature (indirectly)
Fan-pad cooling system	Temperature and humidity
Air conditioners	Temperature and humidity
Shading/thermal screen system	Temperature, light and photoperiod
Fogging/misting system	Temperature and humidity
Heating equipment	Temperature
Humidifiers	Relative humidity and temperature
Fertigation equipment	Moisture content and nutrient status of soil
CO_2 generators	Air composition
Controllers	Operation of all other equipment

It is assumed that a high tech greenhouse would include most of these equipment and controlling systems.

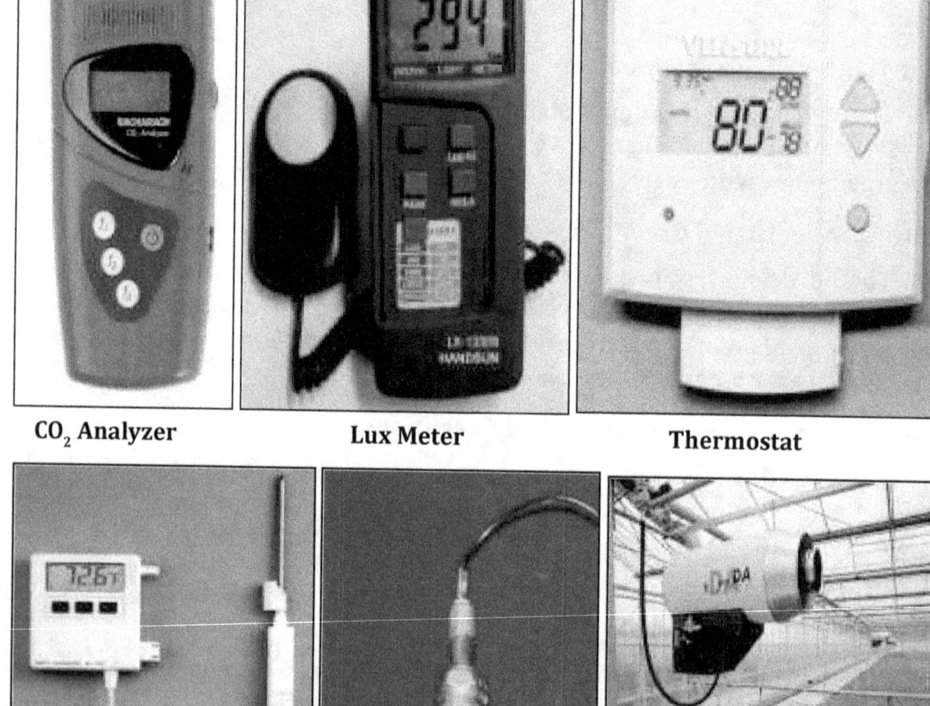

CO$_2$ Analyzer	**Lux Meter**	**Thermostat**
Temperature Sensor	**Oxygen Sensor**	**CO$_2$ Generator**

Figure 2.3. Equipment's Use in Hi-Tech Nursery.

Plant Multiplication under Greenhouse Conditions

Weather fluctuations and rare unpredictable events such as storms, droughts, and floods hinder the propagation of horticultural crops. These factors typically have a detrimental impact on crop output. Greenhouse technology is a contemporary and very efficient method of producing large quantities of high-quality planting material through mass multiplication. Our land offers abundant opportunities for the controlled cultivation of horticulture crops. Continuous propagation throughout the year might be implemented to get more financial gains. An attempt was made to propagate fruit tree nurseries throughout the year in a greenhouse at IARI, New

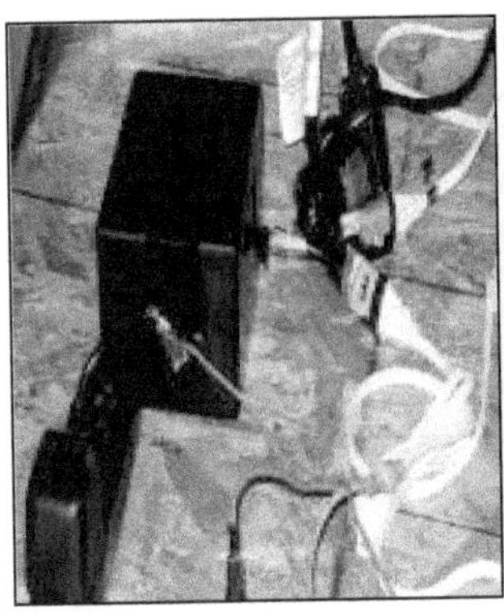 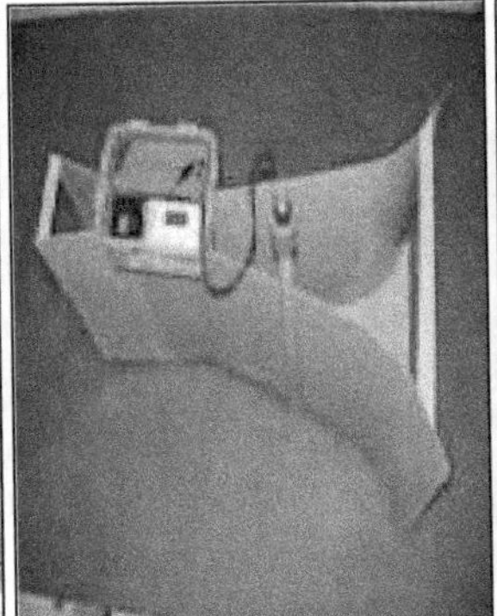 

Figure 2.4. Soil Moisture and Automatic Irrigation Dehumidifier Temperature Sensor Quality Measuring Device.

Delhi. In the months of April-June and July-August, cuttings of kinnow, aonla, ber, and lime were prepared. In the months of September-November, seeds of papaya, mango, and jackfruit were germinated. In February-March, the papaya variety Coorg Honey Dew germinated quite successfully.

Advantages of Plant Multiplication under Greenhouse Conditions

Multiplication of plants under greenhouse conditions offers several distinct advantages over traditional outdoor cultivation methods. Firstly, greenhouses provide a controlled environment where factors such as temperature, humidity, light intensity, and photoperiod can be meticulously regulated. This controlled environment allows for year-round production, irrespective of external weather conditions, thereby extending growing seasons and increasing crop yields. Additionally, greenhouses offer protection from pests, diseases, and harsh weather elements, reducing the need for pesticides and other chemical interventions. The controlled conditions also promote faster and more uniform growth of plants, leading to higher quality produce. Moreover, greenhouse cultivation facilitates efficient water and nutrient management, as these resources can be applied directly to the plants with minimal wastage. Overall, the advantages of plant multiplication under greenhouse conditions include enhanced productivity, improved quality, reduced environmental impact, and greater economic stability for growers. Advantages of Plant Multiplication under Greenhouse Conditions is summarized as under:

- ☆ **Uniformity and Purity of Propagated Plants:** Horticultural crops such as mango, guava, litchi, and papaya exhibit heterozygosity and undergo outcrossing. Conventional seed propagation methods lead to the growth of a vast population of plants that do not possess the desired characteristics. Paradoxically, the enduring character and lengthy development time of these plants pose challenges for early screening. Moreover, by exerting stringent control over edaphic and environmental factors, the process of layering and grafting plantlets can achieve higher success rates and prevent losses caused by seedling mortality. By effectively regulating the temperature of the rooting media and air, it is possible to achieve a rapid production of adventitious roots, resulting in an increased number of propagated plantlets. The flow of cell sap expedites the creation of callus, which subsequently combines to create cambium and the merging of two cell lines (scion and rootstock). Increased vascular connectivity leads to accelerated development and earlier bud emergence. Imposing rigorous regulations on hygienic conditions enables optimal development, consistency, and cleanliness of propagula.

☆ **Genuineness of Planting Material:** The most reliable planting material can be obtained directly from the grower. Therefore, the significance of implementing advanced technology in nurseries for the cultivation of authentic planting material is increasing over time. The grower can fully use the potential benefits of the planting material by understanding the genetic composition of the mother plant. As a result, he can only utilize high-quality mother plants for the purpose of nursery production. The greenhouse buildings allow for rigorous surveillance and control, which is not feasible in open fields. This enables the preservation of the authenticity of planting material.

☆ **Acclimatization of Micro propagated Plantlets:** The process of transferring micro propagated plants from a culture vessel to soil necessitates a gradual hardening phase, which mandates the usage of protected or greenhouse facilities. The control over microclimate ensures the homogeneity of planting material. The plants are shielded from capricious weather conditions and exhibit greater vitality compared to plants cultivated in open field circumstances.

☆ **Economy:** The increasing commercialization and consumeristic nature of society in a globalized economy have had an impact on traditional aspects of nursery and plant propagation. For hi-tech nurseries, the significant initial investment is offset by the substantial productivity achieved by low death rates. Bulk production also reduces the need for fertilization, soil management, irrigation, and other related provisions. At Pantnager, it was observed that the proportion of successful cleft grafting was consistently higher in Polyhouse conditions than to open field conditions, regardless of the time of year. The economic analysis of a low-cost polyhouse for year-round grafting demonstrated a net profit of Rs 39,295 from a 75 m^2 area over a four-month period.

☆ **Disease and Insect-free Planting Material:** In protected cultivation, plants are carefully monitored and treated at appropriate times to prevent disease and pest infestation. The meticulous cultivation of plants from their inception is responsible for the vast supply of robust and thriving plants.

Micro-Irrigation

Micro-watering encompasses low-pressure irrigation systems that utilize spraying, misting, sprinkling, or dripping methods. Micro-irrigation refers to a group of irrigation systems that utilize small devices to distribute water. These devices transport water to the soil surface in close proximity to the plant or directly into the plant's root zone below the soil surface. Nursery producers have modified micro-irrigation systems to meet their need for accurate water distribution. Micro-irrigation is widely employed in the cultivation of plants in greenhouses and nurseries.

Micro-irrigation Systems Components

The components of micro-irrigation consist of pipes, tubes, water emission devices, flow control equipment, installation tools, fittings, and accessories. Irrigation pipeline systems are commonly referred to as branching systems. Branches are assigned designations such as main, sub main, and lateral. It is crucial to select the appropriate size for the main, sub main, and lateral pipes to align with the flow rates from the water source. The essential elements of the system consist of a pump and power unit, a backflow prevention device (if chemicals are employed with water), a filter, a water distribution system, and various devices for regulating water volume and pressure. If the water source originates from a city, municipal, or rural water supply, it is feasible to establish a direct connection. The water distribution system comprises a network of pipes and tubes of varying diameters, ranging from 1/2 inch to 6 inches. A solitary, sizable conduit can transport water from the pump to the periphery of the field. Subordinate pipelines of smaller size can thereafter transport the water to lateral pipelines and ultimately to the emitters. The control components of the control section may consist of a mix of the following devices: pressure regulator, valve, vacuum relief valve, and time clock or controller. Utilize a flow meter to quantify the volume of water. Pressure gauges are used to measure the water pressure at the pump and other specific points. Equipment for injecting fertilizers into the water line is commonly utilized. Backflow avoidance devices are employed to avert the contamination of the water source.

Emission Devices

The emitter is a plastic measuring device that releases a modest yet accurate discharge. The volume of water dispensed by these emitters is often measured in gallons per hour (gph). These emitters reduce water pressure by utilizing extended pathways, narrow openings, or diaphragms. Pressure compensating emitters discharge water at a consistent rate regardless of changes in pressure. Water is delivered by emission devices in three distinct modes: drip, bubbler, and micro-sprinkler.

- ☆ In drip mode, water is applied as droplets or trickles.
- ☆ In bubbler mode, water „bubbles out from the emitters.
- ☆ in micro-sprinkler mode Water is sprinkled, sprayed, or misted.

Drip Irrigation

Bubbler Irrigation

Figure 2.5. Different Micro-Sprinkler Irrigation System.

Drip Irrigation

Water is administered in the form of droplets or trickles in this system. The classification of the drip mode can be further distinguished as either a line source type or a point source type, depending on the placement of the emitters in the plastic polyethylene distribution line. The line source type emitters are placed internally in equally spaced holes or slits made along the line. Water applied from the close and equally spaced holes usually runs along the line and forms a continuous wetting pattern. The point source emitters are externally linked to the lateral pipe. The installer has the option to choose the preferred position based on the planting configuration or evenly distribute them at regular intervals. Water discharged from the emitter at a certain location often creates a circular area that is thoroughly saturated. This is suitable for plants that are widely spaced in orchards, vineyards, or for landscaping trees and shrubs. It is also suitable for row crops that are closely placed in fields and gardens.

Bubbler Irrigation

The water from the bubbler head either cascades down from the emission mechanism or disperses in a radial pattern, resembling an umbrella, covering a few inches of area. The bubbler emitters disperse water pressure by utilizing different diaphragm materials and redirect water through small openings. The majority of bubbler emitters available in the market are advertised as pressure compensating. The bubbler emission devices are supplied with either a single or multiple port outlets. Bubbler heads are mostly utilized in situations where it is desirable to provide deep and concentrated irrigation. The average flow rate from bubbler emitters ranges from 2 to 20 gallons per hour (gph).

Micro-Sprinkler Irrigation

Water is distributed in this system through sprinkling, spraying, or misting. Micro-sprinklers, sometimes referred to as sprinkler or spray heads, are devices that produce water. There exist multiple categories. The emitters function by propelling water through the atmosphere, typically in pre-established configurations. The micro sprinklers are given many names based on the patterns in which they distribute water, such as 1) mini-sprays, 2) micro sprays, 3) jets, or spinners. The sprinkler heads are external emitters that are each individually attached to the lateral pipe. This tube has a diameter ranging from 1/8 inch to 1/4 inch, making it quite compact in size. The sprinkler heads can be affixed to a support post or attached to the supply pipe. The flow rates of micro-sprinkler emitters range from 3 gallons per hour (gph) to 30 gph, depending on the size of the orifice and the pressure in the line.

Advantages of Micro-Irrigation

☆ Micro-irrigation systems provide numerous advantages in hi-tech horticulture, making them a preferred choice for precision water management. One significant benefit is their efficiency in water usage.

These systems deliver water directly to the root zone of plants in controlled amounts, minimizing water loss due to evaporation or runoff. This efficiency not only conserves water resources but also reduces overall irrigation costs.

☆ Furthermore, micro-irrigation systems promote optimal plant growth by ensuring that water and nutrients are distributed evenly and uniformly across the crop area. This uniformity helps to avoid water stress and nutrient deficiencies, thereby enhancing crop health and productivity. The precise delivery of water also reduces weed growth, as only the crop root zone receives irrigation, minimizing competition for resources.

☆ Another advantage of micro-irrigation is its flexibility and adaptability to various crop types and field conditions. These systems can be customized with different emitter types, spacing, and flow rates to accommodate the specific needs of different plants and soil types. Moreover, micro-irrigation systems can be easily integrated with automated control systems and sensors, allowing for real-time monitoring and adjustment of irrigation schedules based on weather conditions and plant requirements.

☆ The advantages of micro-irrigation in hi-tech horticulture include water efficiency, improved crop health and productivity, reduced weed growth, adaptability to diverse conditions, and compatibility with advanced technology for enhanced management and sustainability in agricultural practices.

Hi-Tech Plant Propagation

Hi-tech plant propagation in horticulture represents a significant advancement where technology and science converge to enhance efficiency and quality. Techniques such as tissue culture, micropropagation, and hydroponics allow for the rapid multiplication of plants under controlled conditions. These methods ensure genetic purity, uniformity, and the ability to propagate plants year-round independent of environmental constraints. By utilizing sterile laboratory environments, precise nutrient solutions, and growth regulators, hi-tech propagation optimizes the production of disease-free, high-yield plant varieties. This approach not only accelerates the breeding and selection of desirable traits but also supports sustainable agriculture practices by reducing the reliance on traditional seed propagation and minimizing the environmental impact associated with field cultivation.

Soilless Media

Use soilless media is used in Hi-tech nursery instead of soil base media due to following reasons. Soil base media have Contamination of pathogen, weed seed and Variable composition, CEC and Porosity and it is not easily sterilized.

Function of Media: Media Provide Anchor, moisture, Permit air exchange, Provide nutrients>

Properties: Density, Porosity, Aeration, CEC are the properties of media

Characteristics of Ideal Media

☆ The media must be sufficiently firm to provide anchorage to seed or cuttings It should be decomposed material with high C/N ratio.

☆ Its volumes must be fairly constant when either wet or dry. It should have better water holding capacity (45-65 per cent).

☆ It should be porous (Porosity 55-85 per cent) to drain excess water.

☆ It should be free from weed seeds and harmful pathogens. It should be readily available, reusable and cheaper

☆ Ideal media should have air space 10-30 per cent

Components of Soilless Media

Organic

☆ **Peat Moss:** Peat is formed of the preserved remains of aquatic, marsh, bog, or swamp vegetation that has undergone partial decomposition while submerged in water. The U.S. Bureau of Mines classifies peat into three types: moss peat, reed sedge, and peat humus. Moss peat, commonly known as peat moss in the market, is obtained from sphagnum or other types of moss. The color ranges from a pale tan to a deep brown. It possesses a significant capacity to retain moisture, being able to hold 15 times its weight when dry. Additionally, it exhibits strong acidity with a pH range of 3.2-4.5. It contains a minor proportion of nitrogen, approximately 1 per cent, but lacks substantial amounts of phosphorus or potassium. Reed Sedge peat is composed of the organic matter derived by grasses, reeds, sedges, and other wetland plants such as those found in Florida. Peat exhibits a color spectrum that spans from reddish brown to nearly black. Peat humus is derived from either hypnum moss or reed sedge peat. The hue of the substance ranges from dark brown to black, and it has a relatively poor ability to retain moisture. However, it contains 2.0 to 3.5 per cent nitrogen.

☆ **Sphagnum Peat Moss:** Peat is formed of the preserved remains of aquatic, marsh, bog, or swamp vegetation that has undergone partial decomposition while submerged in water. The U.S. Bureau of Mines classifies peat into three types: moss peat, reed sedge, and peat humus. Moss peat, commonly known as peat moss in the market, is obtained from sphagnum or other types of moss. The color ranges from a pale tan to a deep brown. The substance possesses a significant ability to retain moisture, with a capacity 15 times more than its weight when dry. It exhibits a high level of acidity, with a pH range of 3.2-4.5. Additionally, it contains a minor proportion of nitrogen, approximately 1 per cent, but lacks substantial amounts of phosphorous or potassium. Reed Sedge peat is composed of the organic matter derived from grasses, reeds, sedges, and other vegetation found in swampy areas, such as Florida. Peat exhibits a color spectrum that spans from reddish brown to nearly black. Peat humus is derived from either hypnum moss or reed sedge peat. The hue of the substance ranges from dark brown to black, and it has a limited ability to keep moisture. However, it contains 2.0 to 3.5 per cent nitrogen.

☆ **Pine Bark – Soft Bark:** The process of obtaining pine bark often involves removing it from the trees, grinding it into smaller pieces, and then sorting it into different sizes. Composting bark involves moistening the bark, adding 1 to 2 pounds N/yd3 from either calcium nitrate or ammonium nitrate, forming a pile and then turning the pile every 2 to 4 weeks to ensure proper aeration. Composting bark typically takes 5 to 7 weeks. Aging is a cheaper process, but aged bark has less humus and a greater nitrogen draw- down in the container than composted bark.

☆ **Composted Hard Wood Bark:** The pH of fresh hardwood bark is usually less acid (pH 5 to 5.5) than peat moss or pine bark. Composted bark may be rather alkaline (pH = 7 to 8.5). Hardwood bark typically contains toxic compounds and, for this reason, should be composted before use.

☆ **Sawdust:** These materials are the secondary products of timber mills. They can be effectively utilized in soil blends. Sawdust decomposition provides an ample supply of nitrogen. Typically, it is not advisable to use sawdust or wood products as a container media. The carbon to nitrogen ratio is excessively elevated, necessitating sufficient quantities of nitrogen and composting to prevent adverse impacts on plant growth.

☆ **Coconut Coir:** The raw material, which looks like sphagnum peat but is more granular, is derived from the husk of the coconut fruit. The typical pH range for coir is 5.5 to 6.8, and the average dry bulk density is 4 lbs/ft^3. It contains significant amounts of phosphorus (6 to 60 ppm) and potassium (170 to 600 ppm) and can hold up to nine times its weight in water. Since coir contains more lignin and less cellulose than peat, it is more resistant to microbial breakdown and, therefore, may shrink less. Coir is easier to re-wet after drying than peat moss.

Inorganic Component

☆ **Perlite:** Perlite is a silicaceous mineral that is gray-white in color. It is of volcanic origin and is extracted from lava flows. Perlite has the ability to absorb and retain three to four times its own weight in water. The substance is fundamentally neutral, having a pH range of 6.0 to 8.0, however it lacks any buffering capacity. It lacks cation exchange capability and does not contain any mineral nutrients. Extracted through excessive irrigation. It is highly beneficial for enhancing aeration in the mixture. Perlite, when combined with peat moss, is a widely used substrate for propagating cuttings.

☆ **Vermiculite:** Vermiculite is a mineral with a micaceous structure that undergoes significant expansion when exposed to heat. From a chemical standpoint, it can be described as a compound consisting of magnesium, aluminum, iron, and silicate, with the presence of water molecules. When expanded, Vermiculite has a very low weight (90-150 kg per cubic meter), is chemically neutral, and has high buffering qualities. It is also not soluble in water. It has the capacity to absorb significant amounts of water, specifically 40-54 liters per cubic meter. Vermiculite possesses a comparatively elevated cation exchange capacity, allowing it to retain nutrients for later gradual release. The substance includes magnesium and potassium, however, additional quantities must be obtained from alternative fertilizer sources. Horticultural vermiculite is categorized into four different sizes according on its grade. Particle sizes of No 1 range from 5-8 mm, No 2 is the standard horticultural grade with particle sizes of 2-3 mm, No 3 has particle sizes of 1-2 mm, and No 4, which is particularly suitable for seed germination, has particle sizes ranging from 0.75 to 1 mm.

☆ **Sand:** Sand is composed of fine rock particles, ranging in size from 0.05 to 2.0 mm, which are generated via the process of weathering different types of rocks. The specific mineral content of sand varies depending on the type of rock it originates from. Quartz sand, mostly composed of silica compounds, is commonly employed for propagation purposes. Sand is the densest of the types of soil used for planting, with a volume of one cubic foot weighing around 45 kg (100 lb) when dry. Sand lacks essential mineral nutrients and does not possess buffering capacity or cation exchange capacity (CEC). It is primarily utilized in conjunction with organic substances.

☆ **Rock Wool (Mineral Wool):** This material is derived from several types of rock, such as basalt rock, which is heated to a temperature of approximately 1600°C. As it cools, the rock is transformed into fibers by a spinning process and then compressed into blocks with the addition of a binder. Various kinds of horticultural rock wool are currently available. The material can be found in shredded form, prills (pellets), slabs, bricks, cubes, or blended with peat moss as a combination. Rockwool has a high water-holding capacity while maintaining adequate levels of oxygen.

Soil Sterilization

Steam Pasteurization

Expose steam (212° F) for 30 to 45 minutes. All sections reach at least 180°

F. involves blowing a mixture of steam and air through the media. Aerated steam (140° to 175° F)

Chemical Fumigation

Chemicals methyl bromide and vapam are used commonly for fumigation.

Solarization

Soil and bed condition-Loose and friable soil with no large clods or other debris on the soil surface

Plastic Tarp-Clear, UV-stabilized plastic tarp of 0.5 to 4 mils thick. The tarp material stretch across the soil surface. Using two layers of thin plastic sheeting separated by a thin insulating layer of air increases soil temperatures and the overall effectiveness of a Solarization treatment. The edges of the sheets must be buried to a depth of 5 or 6 inches in the soil to prevent blowing or tearing of the tarp by the wind.

Timing-Long, hot, sunny days are needed to reach the soil temperatures required to kill soil borne pests and weed seed. The longer the soil is heated, the better and deeper the control of all soil pests and weeds will be. 2-week trapping period proved only moderately effective

Micropropagation for Production of Disease-Free Planting Material

Micropropagation is crucial in the cultivation of horticultural crops to ensure the creation of plants that are free from viruses. The majority of horticultural crops are propagated asexually, meaning that when a plant becomes infected with viruses, the disease is passed from one vegetative generation to another. Various techniques are employed to regenerate plants that are free from viruses, including meristem tip culture, nucellar embryogenesis, micro grafting, as well as chemo and thermotherapy.

Methods of Micro Propagation for Production of Disease-Free Planting Material

Meristem Tip Culture

The Meristem is a dome of about 0.1 mm in diameter and 0.25 mm long and protected by developing leaves and scales. The defoliated stem segments are first surface sterilized using good sterilant, *e.g.* ethanol, sodium hypochlorite. The Meristem dome is then dissected under a microscope inside laminar airflow. The exposed meristem tip, which appears as shiny dome, is then severed with the blade and transferred to liquid or solid medium. Murashige and Skoog is most commonly used medium for meristem culture of important horticultural crops due to high concentrations of potassium and ammonium ions and meso-inositol. The pH of the media may be a limiting factor for growth of meristem. The pH should range between 5.5 and 5.8. The technique of Meristem culture has successfully been used to obtain virus-free plants of a number of horticultural crops.

Nucellar Embryogenesis

Nucellar embryogenesis is another technique for mass production of virus-free plantlets in crops like citrus and mango, which are highly polyembryonic in nature. In citrus embryos arise from nucellus or integument adventively which is taken advantage for producing true-to-type plants. In citrus there are some species, which are highly polyembryonic and certain species are monoembryonic. It was found that nucellus taken from fertilized ovules of all monoembryonic cultivars would not develop. Pollination and fertilization are essential for the induction of nucellar embryogenesis even though there are reports of success in induction of embryogenesis using unfertilized ovules. Nucellar embryoids were formed only in polyembryonic Trovita ovule in culture. About 8-10 weeks old ovule of Citrus reticulata have been found as best explant for initiating nucellar embryogenesis under in-vitro condition. MS medium fortified with malt extract and/or paclobutrazol was found to be best for induction of nucellar embryogenesis in citrus. The inorganic salts such as ammonium nitrate, calcium chloride, potassium phosphate and potassium iodide showed significant association with embryogenesis

Thermotherapy

Thermotherapy is particularly beneficial for the treatment of viral and mycoplasmal infections in fruit trees. Meristem cultivation is challenging for fruit trees, making this method highly beneficial. The procedure entails subjecting a branch to a consistent or fluctuating temperature of 37-38 °C for a duration of 20-40 days. Next, the bud that is free from any potential viruses is carefully removed and then attached to a rootstock that is also free from viruses. Thermotherapy, in conjunction with meristem culture, is being implemented on a large scale in the cultivation of strawberries. The precise method by which thermotherapy enables the development of virus-free plants remains unknown. Nevertheless, there is a hypothesis that heat deactivates the virus within the system, hinders the production of viral RNA, and diminishes the movement of viruses.

Micro Grafting or *In vitro* STG

The technique of micro grafting or in vitro shoot tip grafting is being employed to eliminate viruses from significant woody perennial fruit plants. It was initially attempted in citrus. Navarro *et al.,* adapted the method in citrus. Citrus graft-transmissible illnesses caused by viruses, viroids, bacteria, spiroplasms, and phytoplasms result in significant economic losses in the majority of citrus cultivation regions. Their presence leads to a decrease in tree vitality, shortened commercial lifespan, reduced yields, and poor fruit quality. Therefore, they have the potential to become the main limiting factor in production. The previous methods employed, such as Nucellar embryony and Thermotherapy, have proven to be ineffective in this particular situation due to their inherent drawbacks. Therefore, a new approach is needed to successfully regenerate citrus plants that are free from graft-transmissible pathogens and devoid of juvenile characteristics, ensuring the production of healthy trees suitable for commercial plantings.

The Process known as Shoot Tip Grafting

The technique of In Vitro was pioneered by Navarro *et al.* (1975), enabling a 30-50 per cent success rate in grafting, with subsequent transplantation to soil resulting in over 95 per cent survival. The new plant exhibited no juvenile characteristics and the majority of them were devoid of graft-transmissible pathogens.

Techniques for Shoot Tip Grafting

In vitro: It Includes the Following Steps:

- ☆ **Root Stock Preparation:** Seedlings obtained from in vitro seed germination are utilized as rootstock. The seeds are carefully peeled, removing both seed coverings, and then sterilized on the surface. They are then planted in culture tubes measuring 25 × 150 mm, which contain 25 ml of the plant cell culture salt solution developed by Murashig and Skoog in 1962. Consolidated with a 1 per cent concentration The Bacto agar cultures are incubated at a temperature of 27°C under conditions of continuous darkness for a duration of two weeks. The highest rate of successful grafts was achieved using seedlings that were two weeks old. The stem has a height of 3-5 cm and a diameter of 1.6 to 1.8 mm at the grafting point. The Troyer citrange is the predominant rootstock utilized for STG. The following rootstocks have been successfully used for In-vitro STG: Rough lemon, sweet orange, sour orange, Rangpur lime, and Cleopatra mandarin. The rootstock seedling is advised to be placed in a test tube under sterile circumstances and then trimmed, leaving approximately 1.5 cm of the epicotyl. The root is trimmed to a length of 4-6 cm and the cotyledons and axillary buds are eliminated.

- ☆ **Scion Preparation:** Shoot tips can be removed from actively growing vegetative flushes of field or greenhouse plants, as well as from budwood cultivated in vitro. They can also be removed from dormant buds. Field trees are disadvantaged by their seasonal shedding and their slower rate of eliminating some diseases. Greenhouse plants offer the benefit of being able to generate flushes when needed and can undergo heated pretreatment to enhance the effectiveness of disease eradication. The budwood was cultivated in vitro at a constant temperature of 32°C and exposed to 80 µm/m2s illumination for 16 hours daily in the culture medium. Murashige and Skoog is a really beneficial source of shoot tips for STG.

- ☆ **Grafting:** Flushes that are 3 cm long or shorter are gathered, removing any larger leaves, and then trimmed to approximately 1 cm in length.

Afterward, they undergo surface sterilization using a 0.25 per cent solution of sodium hypochlorite. The shoot tip, which consists of the apical meristem and leaf primordia of a tree, is then removed using a razor blade. It typically measures 0.1-0.2 mm in length. The shoot tip is inserted into the incision of the rootstock, aligning its cut surface with the exposed area created by the horizontal cut at the top of the decapitated epicotyle. The size of the leaf primordium ranges from 0.1 to 0.2 mm, which greatly enhances the success rate of grafting, allowing for the highest possible number of successful grafts.

☆ **Culture *In vitro* of Grafting Plants:** Micrografted plants are cultivated in a liquid nutrient media consisting of the Murashige and Skoog plant cell culture salt solution, or the modified Whites solution. This medium is dispersed into test tubes of 25 × 150 mm, with each tube containing 25 ml of the solution. A folded paper platform, with a perforation in its center for inserting the root section of the rootstock, is positioned in the nutritional solution. The cultures are maintained at a constant temperature of 27 °C and receive 16 hours of daily exposure to 40 – 50 µE/ M2S illumination. Five days after grafting, the callus was fully established at the graft union. Vascular differentiation imitation was noted seven days after transplantation. After 11 days of grafting, a full vascular connection was established. Following 4-6 weeks of grafting, the successful grafts are fully developed and can be transferred to the soil.

☆ **Transplanting to Soil:** For a successful graft, it is necessary for the scion to have a minimum of two fully grown leaves before being transferred to the soil. This stage typically occurs 4-6 weeks after the grafting process.

Application of *In-vitro* STG

In-vitro tissue culture, also known as micropropagation, finds extensive application in modern agriculture and horticulture for the rapid multiplication of plants under controlled conditions. This technique involves the growth of plant cells, tissues, or organs in a nutrient-rich medium in sterile laboratory conditions. One significant application of in-vitro tissue culture is the production of disease-free plant material. By starting with sterilized plant tissue, researchers can eliminate pathogens and obtain healthy stock for propagation. Moreover, in-vitro tissue culture allows for the propagation of plants with desirable traits, such as high yield, disease resistance, or specific growth habits. This method facilitates the cloning of genetically identical plants (clones), ensuring uniformity in traits across a large population. This uniformity is particularly advantageous in commercial horticulture for producing consistent quality crops. Additionally, in-vitro tissue culture plays a crucial role in the conservation and preservation of endangered or rare plant species. It offers a controlled environment where plants can be propagated and preserved without the risk of extinction. Furthermore, in horticulture, in-vitro techniques are used for the rapid production of ornamental plants, fruits, and vegetables. It allows for year-round propagation independent of seasonal limitations and climatic variations, thereby ensuring a continuous supply of plant material. The application of in-vitro tissue culture in horticulture provides solutions for disease management, trait selection, conservation efforts, and efficient commercial production, contributing significantly to the advancement and sustainability of modern agricultural practices.

- ☆ Management of graft transmissible pathogens.
- ☆ Sanitation initiatives.
- ☆ Protocols for isolation and restriction of movement.
- ☆ Somatic hybrid regeneration.
- ☆ Plant regeneration from irradiated shoots.
- ☆ Haploid plant regeneration.
- ☆ Development of a genetically stable tetraploid plant with a monoembryonic genotype.
- ☆ Plant regeneration from somaclonal variation tests using adult plant material.
- ☆ Plant regeneration in transgenic organisms.
- ☆ Micrografting techniques have been successfully developed for a variety of fruits including peaches, plums, cherries, apricots, grapes, apples, mandarin oranges, cashews, and tea.

Automation in Hi-tech Horticulture

Automation in hi-tech horticulture represents a transformative shift in the way crops are cultivated, managed, and harvested. By integrating advanced technologies such as robotics, sensors, artificial intelligence, and data analytics, automation enhances precision, efficiency, and sustainability throughout the agricultural process. Automated systems can monitor and control crucial variables such as temperature, humidity, light, and nutrient levels in real-time, optimizing growing conditions for plants. This precision not only improves crop quality and yield but also reduces resource consumption and labor costs. Moreover, automation enables tasks like planting, irrigation, fertilization, and pest management to be carried out with greater accuracy and consistency, freeing up human resources for higher-level decision-making and innovation. As hi-tech horticulture continues to evolve, automation stands at the forefront, driving productivity gains and shaping the future of sustainable agriculture.

- ☆ **Nursery Mechanization:** The essential root medium can be crushed, blended, heated to a specific temperature, and placed in containers or trays. However, the equipment needed for this process must be further developed. The primary medium can be combined in batches by adjusting the concrete mixer. Pasteurization is achieved by subjecting the rooting media to steam at a temperature range of 60-82oC for a specific period

Figure 2.6. Automatic Pot Filling Machine Fill Up, Fertilized and Potted Exactly 4800 Pots per Day.

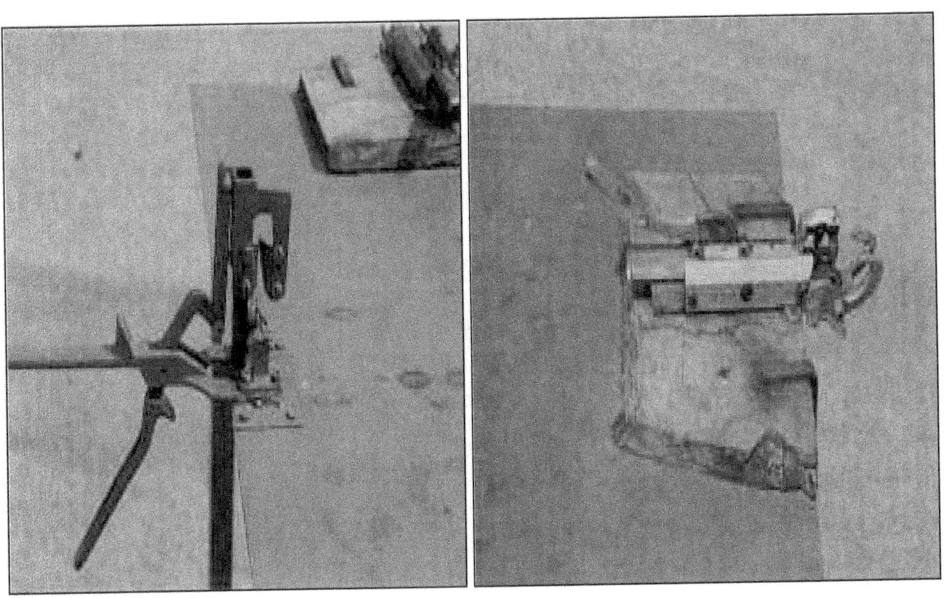

Figure 2.7. Grafting Machines.

of time. Utilize a portable steam generator to produce steam, and then direct the aerated steam through perforated pipes that are buried beneath the media beds. The potting media can be placed in seedling trays or pots for transplanting, using a screw auger. Planters can be used to sow the seeds. A sensor utilizing frequency domain technology, equipped with 4 cm electrodes, was calibrated to accurately measure the volumetric water content and electrical conductivity (EC) in growing material used for horticultural crops.

✰ **Sowing and Transplanting:** The process of transplanting and replacing seedlings in trays incurs significant costs in terms of skilled labour. Trentini designed a computer-controlled equipment for the automated planting of seedlings in boxes, which represents a brief exploration of automation in horticulture. The process involves extracting the seedlings together with a compacted portion of soil from the trays and subsequently relocating them at the desired intervals. It can transplant 2,000 seedlings in a hectare with a single grip, and 5,800 when the actuator is fitted with 3 grippers. A worker manually picking and transplanting seedlings can handle 900-1,200 plants over an 8-hour shift. The process of transplanting nursery plants into the field requires a significant amount of manual labour. Utilizing the method of raising seedlings in block and transplanting enhances the precision and effectiveness of the transplanting process.

✰ **Soil Moisture Measurement:** Accurate and ongoing assessment of soil water content is crucial for interpreting data obtained from field and greenhouse research. This is particularly accurate when it comes to research on the water consumption of plants and its significance in determining the timing of irrigation. There are various devices that can measure moisture, such as the resistance block, tensiometer, and neutron moisture probe. A highly accurate sensor was designed and field tested for six growing seasons to continually monitor leaf thickness in the field. An investigation was conducted to determine the viability of three distinct electrical components as transducers for converting linear changes in leaf thickness into a quantifiable electrical signal. Recent studies have established a strong and direct relationship between the thickness of leaves and their turgor potential, with a correlation coefficient (R2) more than 0.9. This correlation is considered significant and indicates that leaf turgor potential is a reliable and precise indicator of plant-water status and its impact on plant metabolism.

✰ **Irrigation:** The advancement in comprehension of the soil-plant correlation has led to the notion that the most effective use of accessible water resources and optimal plant functioning can be achieved through the avoidance of moisture stress. Seeking data on soil water potential for automated regulation of a micro irrigation system. Granular matrix sensors can be utilized to obtain soil-water potential information. A data-logger needs to be configured to ensure that the soil water potential remains at a constant level by implementing frequent irrigations (up to 8 times per day) using controllers that are attached to solenoid valves. Soil-water potential measurements would offer the essential information to automatically plan frequent drip watering. The feedback enables the preservation of a consistently stable soil water potential within the root zone. By maintaining a consistent soil-water potential in the root zone, it is possible to achieve optimal crop growth while minimizing the risk of leaching. In addition to these factors, significant quantities of valuable irrigation water may be lost as a result of percolation beyond the area where plant roots may access it. The concept of automated irrigation equipment is indeed justified, considering the diminishing water resources and the potential issues of surface and groundwater pollution that may arise from excessive water usage. Farmers should only irrigate when necessary and in the appropriate quantity. Implementing the practice of applying irrigation water at different rates in specific areas of the field will enhance water-use efficiencies. Typically, low lying places have lower water requirements compared to hilltops. However, our current technology provides water evenly across all areas, which should be adjusted to suit the individual needs of each site. Ehlert *et al.,* recommended the utilization of various sensors to gather information on soil factors, weather conditions, crop characteristics, and fertilizer concentrations. This data can then be used to facilitate the automation of micro irrigation systems. Savvas and Adamidis proposed an algorithm for formulating fertilizer solutions depending on soil properties. Benami and Offen (3) proposed fundamental principles for the potential implementation of varying degrees of automation in sprinkler irrigation. Micro irrigation systems have the ability to achieve a very high level of irrigation efficiency, reaching up to 97 percent. In order to get this level of accuracy, it is necessary to employ computers for precise control and automated operation. The main focus will be on enhancing production by precisely applying nutrients and irrigation water based on the crop's physiological growth and the current agroclimatic circumstances. In India, there is a significant lack of knowledge and expertise in the development, operation, and maintenance of automated micro irrigation systems. These technological bundles are purchased at exorbitant prices from countries like Israel and the United States. Indian users are not provided with the technical knowledge in all such cases. Hence, it is imperative to invest in the creation of a native automated micro irrigation system that can provide irrigation and nutrients based on the real-time distribution of soil moisture and concentration of nutrients in the plant's root zone.

☆ **Fertilizer Application:** The technology for the application of fertilizer is adequately advanced, particularly in terms of hardware and software. Currently, the primary aspect that has to be addressed is the establishment of a procedure for formulating a fertilizer suggestion that takes into account the specific characteristics of the soil and crop. Utilizing fertilizers via drip irrigation necessitates the use of specialized fertilizer applicators to ensure precise concentration and application rates in accordance with the water used for irrigation and the requirements of the crops. While venturi devices of ¾ inch size and fertilizer tanks are now accessible in the country, there is a pressing want for more accurate fertigation pumps to effectively administer nutrients in conjunction with irrigation water. The optimal nutrient levels in the soil of the root zone can be achieved by applying the necessary amount of nutrients through irrigation water.

☆ **Insect Pest Management:** Manual sprayers are mostly used for spraying. These procedures are characterized by a high degree of manual labour and can be automated by employing power-operated sprayers. Manual trapping methods are commonly employed for the detection and identification of insect pests. Recent advancements in signal processing and computer technology have made it possible to automatically identify species using various methods, such as image analysis and acoustics. Insects have the ability to produce sound intentionally for communication purposes or unintentionally while eating, flying, or engaging in other movements. This sound can be used for detecting and identifying insects. Researchers at Hull University in the UK are studying methods to automatically identify Orthoptera, which include grasshoppers and crickets, using time domain signal processing and artificial neural networks. A test set consisting of twenty-five species of British Orthoptera has been chosen. The initial findings suggest a categorization rate that is very close to 100 per cent, with an exceptionally low rate of misclassification. This approach has broad applicability to numerous insect pests and other phyla, including birds. The purpose of developing the electronic probe insect counter (EGPIC) system was to create a method for automatically and immediately monitoring insects. The number of insects detected by a network of electronic grain probes placed throughout a storage area is sent to a central computer for display and analysis over time. The EGPIC system enables early detection of nascent infestations, allowing managers to implement targeted control actions as needed, before significant losses occur.

Chapter 3

Micropropagation

Introduction

Micropropagation is a crucial technique used to propagate plants. It involves the controlled growth of plant cells, tissues, or organs in a laboratory setting. Micropropagation encompasses a range of techniques, each with distinct benefits and practical uses. It serves as an excellent substitute for conventional multiplication techniques and guarantees the generation of genetically identical and disease-free plants. This technique is beneficial for cultivating plants that are afflicted with viral or bacterial infections and are unable to generate high-quality, disease-free yields. This includes plants that are incapable of producing seeds, among other examples. This technique is extensively employed in the agriculture and horticulture sectors and is highly effective. Micropropagation is a method of propagating plants by cultivating their cells, tissues, or organs in a controlled laboratory setting. This technique enables the efficient generation of many genetically identical and disease-free plants from a minuscule amount of plant material. It has a crucial role in the preservation of rare and endangered plant species. Micropropagation is extensively employed in several fields such as agriculture, horticulture, forestry, and scientific research. Micropropagation is a significant method for plant reproduction. It entails a distinctive and effective method for the large-scale creation of plants with specific characteristics. It has a crucial function in agriculture, horticulture, forestry, and medicines, since it contributes to enhanced productivity, genetic homogeneity, and the preservation of plant biodiversity. Micropropagation is an important technique in plant breeding and conservation due to its capacity to generate a significant quantity of genetically identical and disease-free seedlings. The efficacy and regulated process of micropropagation is beneficial for the large-scale production of plants. Micropropagation is crucial for the preservation of rare and endangered plant species. Therefore, it may be inferred that this is a highly significant biological technique for humans.

History

In 1902, a German physiologist, Gottlieb Haberlandt for the first time attempted to culture isolated single palisade cells from leaves in knop's salt solution enriched with sucrose. The cells remained alive for up to one month, increased in size, accumulated starch but failed to divide. First attempt by Haberlandt (1902) grew palisade cells from leaves of various plants but they did not divide continuously growing culture of meristematic cells of tomato on medium containing salt yeast extract and source and 3 Vit. B (pyridoxine, thiamine, nicotinic acid) established the importance of additives. Miller and Skoog (1953), University of Wisconsin–Madison discovered kinetin, a cytokine that plays an active role on organogenesis.

Example

A few of the examples of Micropropagation are as follows:

☆ Plants such as palm, plantain, banana, jojoba, pineapple, rubber tree, cassava, yam, sweet potato, tomato, pine, date, brinjal, *etc.* have been produced using this method.

☆ An example of micropropagation is the commercial production of different orchids. This technique has been exploited for the production of different types of orchids worldwide.

Stages of Micropropagation

The process of Micropropagation is a multi-step process. It consists of five steps which are Stage 0, Stage 1, Stage 2, Stage 3, and Stage 4. Each step has its specific requirements along with the problems. These are elaborated as follows:

☆ **Stage 0: Selection of an Explant:** This is the initial phase of the micropropagation process. The stock plants are chosen and grown under controlled conditions before being used for culture initiation.

☆ **Stage 1: Initiation of Cultures:** This stage focuses on the establishment of aseptic cultures of the plant which is to be micropropagated by using suitable explants. If at this stage the explant selected is sterilized properly then the success rate of the procedure increases. This is a very important stage.

☆ **Stage 2: Multiplication:** This process has to be efficient for the success of micropropagation. Shoot multiplication happens by these three methods — regeneration from callus, forced axillary branching, and direct adventitious bud formation from the explant.

☆ **Stage 3: Shoot Elongation and Rooting:** Somatic embryos have both shoot and root primordia and can thus result in complete plants. However, the shoots that are formed by the processes mentioned in Stage 2 require an additional step of rooting for complete plant formation.

☆ **Stage 4: Transplantation and Acclimatization:** Finally, the plantlets formed are transferred to the soil or potting mixture. This is a very critical step as the plants exposed to the in vitro environment though grow well but suffer from many morphological, anatomical, cytological, and physical abnormalities. Due to this, it is necessary to be very careful during this stage (Figure 3.1).

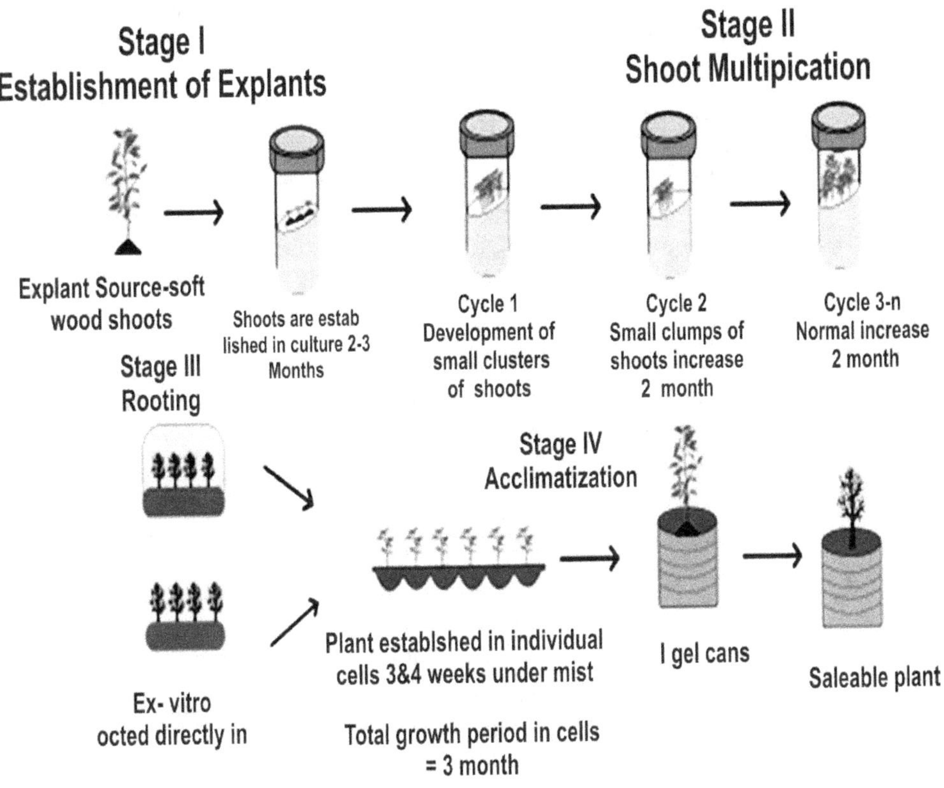

Figure 3.1. Stages of Micropropagation.

Methods of Micropropagation

Micropropagation encompasses a range of methods used to efficiently propagate plants in a controlled environment, allowing for large-scale production. The choice of procedures is contingent upon the particular plant species, kind of tissue, and the intended results.

☆ **Meristem Culture:** This technique involves the isolation and cultivation of the meristematic tissue, found at the shoot tip or the base of the plant. Meristem culture is crucial for obtaining virus-free plants.

☆ **Callus Method:** In this method, plant tissues, such as leaves or stem segments, are cultivated to form callus. The callus is then manipulated to regenerate shoots and roots, facilitating the production of multiple plantlets.

☆ **Suspension Culture Method:** Suspension culture involves placing small tissue fragments or cells in a liquid medium. This technique is used when a homogeneous distribution of cells is desired. It is used for large-scale micropropagation.

☆ **Embryo Culture:** In this technique, immature embryos are isolated from seeds and cultured to develop into plants. It is used in the propagation of certain orchid species and other plants.

☆ **Protoplast Culture:** Protoplasts are plant cells that have been isolated and had their cell membranes removed. Subsequently, these cells are cultivated in a nutrient-rich solution to facilitate the regrowth of their cell walls and the development of whole plants. Protoplast cultivation is advantageous for species that have had limited success with other approaches.

Key Points of Micropropagation

Micropropagation encompasses various crucial factors that contribute to the efficacy of the process. The following are key elements of micropropagation:

☆ **Explants Selection:** It involves careful selection of suitable explants, such as shoot tips, leaves, or meristems as it determines the success of further stages.

☆ **Hormonal Control:** Plant hormones, such as auxins and cytokinins, play an important role in regulating cell division. Precise control of hormones is essential for shoot initiation or root development.

☆ **Transfer to Soil:** Rooted plantlets are transferred to soil or other growing media for continued development. This marks the transition where plants grow in a more natural environment.

☆ **Quality Control:** Regular monitoring and quality control measures, including inspections and diagnostic tests, ensure the production of healthy and disease-free plants.

Applications of Micropropagation

Micropropagation, also known as tissue culture, is a valuable biotechnological tool used in agriculture, horticulture, and forestry. It involves the propagation of plants from small plant parts under sterile conditions, typically using nutrient media. This technique allows for the rapid multiplication of plants with desirable traits, such as disease resistance, high yield, or unique aesthetics. Micropropagation is extensively used in the mass production of elite cultivars, conservation of rare and endangered species, and the propagation of plants that are difficult to propagate by conventional methods. It also plays a crucial role in the production of disease-free planting material, ensuring healthier crops and contributing to sustainable agriculture practices globally.

The application of micropropagation are as follows:

- ☆ It is used for growing commercially important plants in the whole world. By this method, over a billion plants are grown annually.

- ☆ Plantlets produced due to micropropagation can be grown anytime in the year.

- ☆ The main advantage of this procedure is that the cultures can be formed using very small parts of plants and its scale-up multiplication can be done in a short period of time within limited space.

- ☆ It can be used for the propagation of diseased plants. For this, the explant is taken from virus-free meristematic tissues.

Advantages of Micropropagation

Micropropagation offers several distinct advantages that make it a powerful tool in modern agriculture and horticulture. Firstly, one of its primary benefits is the rapid production of large numbers of genetically identical plants from a small amount of initial plant material. This efficiency is particularly advantageous for propagating plants with desirable traits, such as high yield, disease resistance, or unique aesthetic qualities, ensuring consistency in crop or ornamental plant production. Secondly, micropropagation enables the propagation of plants that are difficult or slow to propagate through traditional methods like seeds or cuttings. This includes plants with low seed viability, sterile hybrids, or those that do not root easily from cuttings. By bypassing these limitations, micropropagation expands the range of species and cultivars that can be commercially produced. Another significant advantage is the production of disease-free plants. By starting with sterilized plant material and maintaining strict sterile conditions throughout the process, micropropagation reduces the risk of transmitting diseases and pathogens that may affect conventional propagation methods. This is crucial for maintaining healthy plant populations and preventing the spread of diseases in agricultural and natural ecosystems. Additionally, micropropagation allows for the conservation and preservation of rare and endangered plant species. By culturing plant tissues in vitro, conservationists can propagate and reintroduce threatened species into their natural habitats or maintain them in botanic gardens and seed banks, contributing to biodiversity conservation efforts globally. Moreover, micropropagation offers precise control over the growth conditions of plants, such as nutrient levels, hormones, and environmental factors like light and temperature. This control facilitates the optimization of growth rates and the development of plants under conditions tailored to their specific needs, further enhancing productivity and quality. Micropropagation is a versatile and efficient technique with broad applications in agriculture, horticulture, and conservation. Its ability to rapidly produce genetically uniform, disease-free plants from limited initial material makes it an indispensable tool for sustainable crop production, biodiversity conservation, and the preservation of rare and valuable plant species.

The advantages of micropropagation are as follows:

☆ **Rapid Multiplication:** Micropropagation allows the rapid multiplication of plants. Thus, it is a quick and efficient method for the production of a large number of genetically identical plants in a short time.

☆ **Disease-Free Plants:** The sterile conditions of tissue culture prevent the transmission of diseases. So disease-free plants can be produced using it.

☆ **Preservation of Rare Species:** It has an important part in the conservation of rare plant species because this technique provides a controlled environment for plant propagation. Thus, rare species can be conserved using it.

Disadvantages of Micropropagation

While micropropagation offers significant advantages, it also comes with several disadvantages that need consideration. One major drawback is the potential for genetic uniformity. Because micropropagation produces clones of the original plant, there is little genetic variation among the propagated plants. This lack of diversity can increase susceptibility to pests, diseases, and environmental stresses. In agriculture, this uniformity may limit resilience to changing environmental conditions or evolving pathogens, leading to increased crop vulnerability. Another concern is the potential for somaclonal variation. During the tissue culture process, genetic changes can occur due to somatic mutations or epigenetic modifications. These variations may not always be beneficial and can lead to unpredictable changes in plant characteristics, such as altered growth patterns, reduced vigor, or changes in biochemical composition. Managing and minimizing somaclonal variation is a significant challenge in micropropagation.

Additionally, micropropagation requires specialized skills, equipment, and facilities. The process is labor-intensive and requires strict aseptic conditions to prevent contamination. Establishing and maintaining sterile conditions throughout the tissue culture process can be costly and technically demanding, especially for small-scale producers or in regions with limited resources. Furthermore, the initial establishment of plant cultures through micropropagation can be time-consuming and expensive. It often involves the use of growth regulators, hormones, and nutrient media tailored to the specific needs of each plant species, which adds to production costs. The high initial investment in infrastructure and expertise may not always be feasible for all growers or conservation efforts. Ethical considerations also arise, particularly concerning the intellectual property rights associated with genetically identical clones. Issues related to ownership, distribution, and commercialization of micropropagated plants can be contentious, especially in the context of agricultural biotechnology and plant breeding. Micropropagation offers valuable benefits such as rapid propagation, disease-free plants, and conservation of endangered species, it is important to consider its limitations and challenges. Addressing issues such as genetic uniformity, somaclonal variation, cost-effectiveness, and ethical concerns is crucial for maximizing the benefits of micropropagation while minimizing its drawbacks in various applications.

The disadvantages of Micropropagation are as follows

- ☆ **Genetic Uniformity:** The genetic uniformity that is an advantage can also be a disadvantage. As micropropagation produces genetically identical plants if a pest or disease affects one plant, it is likely to affect all plants.

- ☆ **Loss of Genetic Diversity:** Micropropagation results in genetically identical plants, limiting genetic diversity within the propagated population. This reduction in diversity may limit the adaptability of the plants to changing environmental conditions which may lead to loss in genetic diversity.

- ☆ **Sensitivity to Environmental Conditions:** Plants grown through micropropagation are more likely to be sensitive to changes in environmental conditions. Variations such as temperature, light intensity, or nutrient concentrations can affect the success of the propagation process.

Procedure of Micropropagation Practice

Establishment

The process of micropropagation commences with the careful selection of the plant material that will be propagated. Pristine stock materials devoid of viruses and fungi play a crucial role in the cultivation of the most robust plants. After selecting the plant material for cultivation, the process of collecting the explant commences, which varies according on the specific type of tissue to be utilized. This can include stem ends, anthers, petals, pollen, and other plant tissues. The explant material is subsequently subjected to surface sterilization, typically involving numerous rounds of bleach and alcohol washes, followed by a final rinse in sterilized water. A minute fragment of plant tissue, occasionally consisting of only one cell, is placed onto a growth medium that usually contains sucrose as an energy source and one or more plant growth regulators. Typically, agar is added to the medium to thicken it and form a gel that provides support for the explant during its growth. Certain plants can be cultivated on basic substrates, while others necessitate more intricate substrates to thrive; the growth and specialization of plant tissue is contingent upon the composition of the substrate. Media containing cytokinin are utilized to induce the formation of branching shoots from plant buds.

Multiplication

Multiplication refers to the process of amplifying the quantity of tissue samples generated during the initial stage. After the successful initiation and expansion of plant tissue, the establishment stage is followed by multiplication. By undergoing multiple iterations of this procedure, a solitary explant specimen can be multiplied from one to numerous or even countless plants. The procedures and media used for multiplication can vary depending on the type of tissue being generated. If the plant material cultivated consists of callus tissue, it can be fragmented into smaller pieces using a blender and subsequently re-cultivated on the same type of culture media to promote further growth of callus tissue. If the tissue is cultivated as diminutive plants known as plantlets, hormones are frequently supplemented to induce the production of numerous little offshoots, which can be detached and subjected to further cultivation.

Root Formation

The plantlets were transplanted from the plant media to the soil, which was enriched with vermicompost. This method is conducted to facilitate the adaptation of plantlets to the soil, as they were previously cultivated in a plant medium. Once the plantlets have reached a certain stage of growth, they are relocated to the field. This phase entails the application of certain treatments to the plantlets/shoots to stimulate root development and enhance their resilience. The procedure is conducted in vitro, which means it takes place in a sterile setting resembling a "test tube". Hardening involves the process of preparing plants to adapt to their natural growth environment. Up until this point, the plantlets have been cultivated in optimal settings specifically intended to promote fast growth. Because the plantlets are carefully managed during their growth, they frequently lack fully operational outer coverings. As a result, they are extremely vulnerable to illness and ineffective in their utilization of water and energy. Under in vitro circumstances, characterized by high humidity, plants frequently fail to develop a functional cuticle and stomata, which are essential for preventing excessive water loss. Upon removal from their culture, the plantlets require a period of acclimation to adapt to the surrounding natural environmental conditions. Hardening often entails gradually acclimating the plantlets from a high-humidity, low light, and warm setting to a standard growing environment suitable for the specific species.

Acclimatization

During the last phase of plant micropropagation, the plantlets are extracted from the plant media and relocated to soil or, more frequently, potting compost. This allows them to continue growing via traditional methods.

Plant Tissue Culture Media Preparation

The concept of Plant Tissue Culture originated from the ideas proposed by the German Scientist, Haberlandt, at the start of the 20th century. This was the initial phase of tissue culture, followed by the commercialization of the technology in the 1970s. At present, the world is focused on the predicted future scarcity of grain, the greenhouse effect, and the overall environmental imbalance. Titan Media has introduced a wide range of Ready to use PTC Medium, taking into consideration various factors and in order to support humanity. The PTC Medium includes ingredients such as Plant Growth Regulators or Phytohormones (Auxins, Cytokinin, Gibberellins, Abscisic and others), Macro and Micro Nutrients (Nitrogen, Potassium, Phosphorus, Sulphur, Magnesium, Calcium, Boron, Zinc, Iron, Iodine, Vitamins, Amino acids, Carbohydrates and others), Gelling Agents (Gelrite, Agar-Agar and others), Adsorbing Agents, Buffering Agents, Pure Antibiotics, and a range of other products used in the process. Plant tissue culture media preparation relies on the distinctive characteristic of cell totipotency. Cell totipotency refers to the capacity of a plant cell to undergo regeneration and develop into an entire plant. During this procedure, the removed bud is put into a tube that contains a sterile nutrient medium. The efficacy of tissue culture is highly contingent upon the choice of explant stage, duration of sterilization, and the specific culture media employed. Distinct plant species necessitate distinct compositions of culture media. Plant tissues are cultivated in vitro using synthetic media that provide the essential nutrients required for their growth. The efficacy of plant cultivation as a method of plant multiplication is significantly impacted by the characteristics of the culture media employed. Rich tissue culture media serves as a favorable nutrition supply for bacteria and fungus, so it is necessary to implement measures to prevent microbial contamination in all in vitro treatments.

Tissue culture media used for in vitro cultivation of plant cells are composed of following basic components: The fundamental constituents employed in tissue culture media include a complex blend of inorganic nutrients such as essential elements or mineral ions, organic supplements like vitamins and/or amino acids, a carbon source like sucrose sugar, gelling agents or solidifying agents, plant growth regulators, growth promoters, complex organics, activated charcoal, pH regulators, and antibiotics.

Complex mixture of salts is Inorganic nutrients (both micro- and macro-elements: C, H, O, N, P, S, Ca, K, Mg, Fe, Mn, Cu, Zn, B, Mb). For healthy and vigorous growth, intact plants need to take up from soil: relatively large amounts of some inorganic elements (the major plant nutrients): ions of Nitrogen (N), Potassium (K), Calcium (Ca), Phosphorous (P), Magnesium (Mg) and Sulphur (S). And some other elements in small quantities (minor plant nutrients or trace elements): Iron (Fe), Nickel (Ni), Chlorine (Cl), Manganese (Mn), Zinc (Zn), Boron (B), Copper (Cu), and Molybdenum (Mo). Among the micronutrients, iron requirement is very critical. Chelated form of Iron and Copper are commonly used in culture media.

Organic supplement: Vitamins and amino acids, such as Nicotinic acid, Thiamine, Pyridoxine, and myo-Inositol (Vitamin B), as well as Arginine, are necessary for the growth of plant cells in a laboratory setting. The plant cells in culture have the capacity to produce vitamins, similar to natural plants. However, the number of vitamins synthesized is insufficient to facilitate optimal cell proliferation in culture. Consequently, the medium is enriched with vitamins to promote optimal cellular growth. Likewise, amino acids are introduced into the cell cultures to promote cell proliferation and establish the cell lines. Glycine is the most frequently utilized amino acid. Additionally, Asparagine, Aspartic acid, Alanine, Glutamic acid, Glutamine, and Proline are also utilized. Amino acids serve as a reservoir of nitrogen in its reduced form and, similar to ammonium ions, their absorption leads to acidification of the surrounding environment. Organic acids, particularly the intermediates of the Kreb's cycle such as Citrate, Malate, Succinate, and Pyruvate, have a positive impact on the proliferation of plant cells. Occasionally, antibiotics such as Streptomycin and Kanamycin are used in the medium to inhibit the proliferation of microbes.

Carbon sources: Typically, sugar is used. Sucrose, often known as table sugar, is a type of carbohydrate. In the context of plant cells and tissues in a culture medium, they are heterotrophic, meaning they rely on external sources of carbon for energy. During the sterilizing process of the medium, sucrose undergoes hydrolysis to become glucose and fructose. The plant cells then preferentially consume glucose first, followed by fructose. Other carbohydrates, including as lactose, maltose, and galactose, have been utilized in culture medium, albeit with restricted efficacy.

Gelling or Solidifying agents like Agar: Plant tissue culture media can be utilized in either liquid or solid states, depending on the specific type of culture being cultivated. Typically, a gelling agent called Agar, which is a polymer derived from red algae (specifically Gelidium amansii), is used to solidify the liquid media. The agar derived from seaweeds offers a solid substrate for cell growth, preventing suffocation and death caused by submersion in a liquid media devoid of oxygen. Cultures of cells are cultivated in a liquid media without agar, and these cultures are regularly aerated either by bubbling sterile air or by gently agitating the medium. Additional infrequently employed solidifying agents include Gelrite, Biogel (polyacrylamide pellets), Phytagel, and pure Agarose, as well as many types of Gellan gums.

Plant Growth regulators (*e.g.* Auxins, Cytokinins and Gibberellins): Plant hormones are crucial for the growth and differentiation of cultured cells and tissues. The several types of plant growth regulators used in culture media include: Auxins, Cytokinins, Gibberellins, Abscisic acid, Ethylene, 6 BAP (6 Benzyladenine), I.A.A (Indole Acetic Acid), I.B.A. (Indole 3- Butyric Acid), Zeatin, and Zeatin Riboside. Auxins promote cell division and root differentiation. Auxins stimulate cellular proliferation, cellular elongation, and the development of callus in cultures. For example, 2,4-dichlorophenoxy acetic acid is frequently included as an auxin in plant cell cultures.

☆ The Cytokinins stimulate the process of cell division and differentiation. Cytokinins boost RNA production and enhance protein and enzyme activity in tissues. Kinetin and benzyl-aminopurine are the predominant Cytokinins utilized in plant cell cultures. Gibberellins are mostly utilized to stimulate the growth of plantlets from adventive embryos generated in culture.

☆ Abscisic acid is employed in plant tissue culture to facilitate specific developmental pathways, such as somatic embryogenesis. Abscisic acid (ABA) suppresses cellular proliferation.

☆ Ethylene plays a role in regulating the ripening process of climacteric fruits, however its application in plant tissue culture is not widely adopted.

☆ Certain plant cell cultures have the ability to generate ethylene, which, when it accumulates to a significant extent, can impede the growth and progression of the culture. The ratio of auxins and cytokinins in plant tissue culture media is crucial for the morphogenesis of culture systems.

Embryogenesis, callus initiation, and root initiation occur when the ratio of auxins to cytokinins is elevated. In order to promote axillary proliferation and shoot proliferation, it is necessary to maintain a low ratio of auxins to cytokinins. Gibberellic acid (GA3) stimulates the proliferation of callus tissue and promotes the elongation of dwarf plantlets.

Some examples of plant growth boosters include Naphthalene acetic acid and Humic acid (in granular or powder form). Plant growth promoters (PGP) are chemicals utilized to optimize nutrient utilization and enhance plant development. Oftentimes, certain crops fail to achieve optimal yields despite receiving enough nutrition input. Physiological inefficiency in plants is the cause of these impacts. Consequently, these plant growth regulators (PGA) have a significant impact on seed germination, fruit ripening, nutrient absorption, protein synthesis, immune response, stress tolerance, reduction in flower and fruit loss, and overall plant development improvement.

Specific needs may warrant the addition of complex organics such as Casein hydrolysate, Coconut milk, Malt extract, Yeast extract, Tomato juice, *etc.* Casein hydrolysate has demonstrated notable efficacy in tissue culture, while potato extract has been identified as beneficial for anther cultivation. Nevertheless, these natural extracts are intentionally avoided due to their uncertain composition, which not only varies from batch to batch but also changes with time, hence compromising the repeatability of experimental outcomes.

Activated charcoal has a dual effect on the growth of plant cultures, either promoting or inhibiting it, depending on the specific plant species being cultivated. According to reports, it has been found to promote development and specialization in orchids, carrots, ivy, and tomatoes, whereas it hinders the growth of tobacco and soybeans. It has the ability to absorb brown-black pigments and oxidized phenolics that are formed during culture, hence reducing toxicity. In addition, it has the capacity to absorb various chemical molecules such as PGRs and vitamins, which can potentially hinder growth. Activated charcoal has the ability to darken the media, which in turn promotes root development and growth.

The pH level has an impact on the absorption of ions as well as the solidification of the gelling agent. The ideal pH for culture media is 5.8 prior to sterilization. Growth and development in vitro are significantly inhibited by pH values below 4.5 or over 7.0. The pH of culture media typically decreases by 0.3 to 0.5 units after autoclaving and continues to fluctuate during the culture period due to oxidation and the differential absorption and release of chemicals by the growing tissue. Plant cell growth ceases in the medium when the pH exceeds 7.0 or falls below 4.5.

Antibiotics Antimicrobial compounds are employed in plant tissue culture media. The efficacy of combining antibiotics and chemical biocides in inhibiting microbial contamination in plant cultures has been evaluated. Phillips (1981) conducted studies on six antibiotics, namely Benzyl penicillin, Phosphomycin, Chloramphenicol, Streptomycin, Rifampicin, and Nalidixic acid, to assess their effectiveness in preventing bacterial infection in cultures of Jerusalem artichoke (Helianthus tuberosus). Rifampicin was the sole substance capable of managing bacterial contamination without impacting the rates of plant cell division, cell differentiation, or DNA synthesis in cultivated explants. Nevertheless, Rifampicin induced a rise in protein synthesis.

Antibiotics aid in reducing bacterial infections in plant cell and tissue culture. Additionally, it aids in the inhibition of mold and yeast infections in cell cultures. Whereas, it aids in the eradication of Agrobacterium species following the modification of plant tissue. Agrobacterium's capacity to transmit genes to plants and fungus is harnessed in the field of biotechnology, namely in genetic engineering to enhance plant traits. The genes intended for insertion into the plant are replicated and inserted into a plant transformation vector, which includes the T-DNA section of the disarmed plasmid. Additionally, a selectable marker, such as antibiotic resistance, is included in the vector to facilitate the identification and selection of successfully transformed plants. Plants are cultivated on antibiotic-containing medium after undergoing transformation. Plants lacking integration of the T-DNA into their genome will perish. Agroinfiltration is an alternate way.

The aforementioned elements, along with Carbon (C), Oxygen (O), and Hydrogen (H), are the 17 essential elements. While certain elements like Cobalt (Co), Aluminium (Al), Sodium (Na), and Iodine (I) are considered necessary or advantageous for certain species, their overall importance has yet to be fully confirmed.

Epstein (1971) defines essential ingredients for plant growth as those that are necessary for a plant to complete its life cycle.

    ☆ Its function is highly specialized and cannot be fully substituted by any other element.

- ☆ Its impact on the organism is immediate and not mediated by the environment.
- ☆ It is a component of a molecule that is recognized to be indispensable.

Plant Tissue Culture Medium Preparation

One liter of growth medium is sufficient to prepare approximately 100 growing tubes.

- ☆ Measure the necessary quantity of Murashige and Skoog Basal Salts with minimal organics medium (MSMO).
- ☆ To prepare 1 liter of Murashige and Skoog Basal Salts with low organics medium, it is recommended to use a 2-liter container due to the medium's tendency to boil up.
- ☆ Combine the desiccated medium with the water and agitate for one minute until it is fully dissolved.
- ☆ Incorporate the preferred thermally resistant additives into the solution. The user's text consists of a bullet point symbol. Measure out 6 grams of Agar.
- ☆ Incorporate it into the MSMO solution.
- ☆ Apply gentle heat to the solution while stirring until complete dissolution of the agar is achieved.
- ☆ Increase the amount of distilled water until the total volume reaches 1 liter.
- ☆ Transfer the medium, which is still warm, into the polycarbonate tubes until it reaches a depth of approximately 15 mm.
- ☆ Insert the tubes (with loosely placed lids) into an autoclave and subject them to sterilization for a duration of 20 minutes. (Please be aware that elevated temperatures can lead to suboptimal cellular proliferation).
- ☆ Allow the autoclave to cool down, then extract the tubes and securely fasten the lids.
- ☆ Incorporate thermolabile additives into the mixture after autoclaving.

Titan Media Offers Ready to Use Plant Tissue Culture Media

- ☆ Banana Micropropagation Medium
- ☆ Banana Multiplication Medium
- ☆ Gamborg Medium
- ☆ Knudson-C Orchid Medium

The MS medium, introduced by Murashige and Skoog in 1962, is the most widely utilized nutritional medium.

Titan Media Offers Plant Tissue Culture Media Ingredients

- ☆ Macro and Micro Nutrients: Ammonium chloride, Ammonium molybdate, Ammonium sulphate, Potassium nitrate, Potassium sulphate, Sodium nitrate, Urea, Zinc nitrate and many more.

- ☆ Carbohydrates: Dextrose, Fructose, Maltose, D (+) Mannose, D- Ribose, D- Sorbitol, Sucrose, Tomato powder and many more.

- ☆ Amino acids: Aspartic acid, Glycine, L- Lysine mono- HCl, L- Proline, L- Tyrosine

- ☆ Plant Growth Regulators: IAA, IBA, NAA, 2,4,5 -T

- ☆ Antibiotics: Amicillin Sodium, Ciprofloxacin, Kanamycin sulphate, Gentamycin, Polymixin, B sulphate, Vancomycin

- ☆ Vitamins-L: Ascorbic acid, D (+) Biotin, Niacin, Riboflavin, Retinol acetate, Thiamine HCl

- ☆ Gelling Agents: Agar Agar, Agarose, Gelrite, Gellan Gum

- ☆ Organic acids: Citric acid, Casein enzymatic hydrolysate, Malt extract, Soya peptone, Yeast extract and many more.

- ☆ Buffering agents: ACES buffers, CHAPS buffers, HEPES buffers, MES buffers.

Methods for Plant Tissue Culture

- ☆ Embryo culture refers to the aseptic isolation and cultivation of an immature or mature zygotic embryo in vitro, with the objective of creating a viable plant.

- ☆ Meristem culture refers to the aseptic isolation and cultivation of shoot tips or meristems in a controlled environment, typically in a laboratory setting. The purpose of this technique is to generate genetically identical copies of plants, free from viral infections, or to preserve the genetic material (germplasm) of the plant by cryopreservation.

- ☆ Callus culture refers to the aseptic initiation and proliferation of callus in a laboratory setting, with the objective of getting enhanced plants or secondary plant products.

- ☆ Anther culture refers to the process of isolating the anther and cultivating a haploid callus from the pollen in a sterile environment.

- ☆ Protoplast culture is an advanced technique that goes beyond cell suspension culture. In this process, the cell walls of suspended cells are eliminated using enzymes that break down the cellulose. This results in isolated protoplasts, which are cells enclosed by a semi-permeable membrane.

- ☆ Cell Suspension: A cell suspension is a product derived from callus culture, where callus refers to a cluster of undifferentiated cells. When these cells are separated and cultured in a liquid medium, they form a cell suspension.

☆ Fern Spore culture: Cultivating fern spores in vitro is not a traditional tissue culture method, but rather a technique for cultivating spores in a controlled and sterile environment.

$$\overline{\textbf{\textit{Chapter 4}}}$$

Hi-Tech Field Preparation and Planting Methods

Hi-Tech Field Preparation

Hi-tech field preparation involves meticulous planning and execution to create a conducive environment for advanced technological activities. Beginning with site selection, thorough surveys assess terrain, soil composition, and environmental factors to ensure suitability. Infrastructure planning follows, determining needs for robust utilities like electricity, water, and telecommunications, crucial for powering and connecting sophisticated equipment. Ground preparation includes clearing debris and grading land for stability, essential for constructing specialized facilities and platforms. Rigorous construction and installation procedures adhere strictly to technical specifications and safety standards, ensuring optimal functionality of installed systems. Security measures are paramount, safeguarding against unauthorized access and potential threats to sensitive equipment and data. Environmental considerations play a pivotal role, requiring adherence to regulations and implementation of sustainable practices to minimize ecological impact. Testing and commissioning phases validate system integrity, confirming operational readiness before formal handover to operational teams. Comprehensive documentation, including plans and operational manuals, supports ongoing maintenance and upkeep, essential for sustaining peak performance and longevity of hi-tech infrastructure. Hi-tech field preparation integrates cutting-edge technology with meticulous planning and execution, aiming to establish a secure, efficient, and sustainable environment for advanced technological endeavours.

"Hi-tech Field Preparation" likely refers to preparing a field or area for activities involving high technology, which could encompass various steps depending on the specific context. Here's a general outline of what might be involved:

Site Selection and Surveying

In the intricate realm of hi-tech agriculture, where precision and efficiency are paramount, site selection and surveying form the bedrock upon which successful ventures are built. The meticulous process begins with a discerning eye cast over potential locations, where every contour of the land and nuance of the environment is carefully evaluated. First and foremost, climate reigns supreme in the considerations for site selection. The optimal balance of temperature, humidity, and sunlight exposure must be identified to cultivate ideal conditions for crops. Whether it be the sheltered embrace of a valley or the elevated perch of a hillside, each locale offers a distinct microclimate that can significantly impact crop health and yield.

Equally pivotal is the accessibility and availability of essential resources. Water, a lifeblood for any agricultural endeavor, demands meticulous assessment for both quality and quantity. Access to reliable utilities, such as electricity for powering advanced technologies and telecommunications for seamless connectivity, is also non-negotiable in the modern agricultural landscape. Beneath the surface lies the silent protagonist of soil quality and composition. Through comprehensive soil surveys, the fertility, texture, and nutrient levels are meticulously scrutinized. The very foundation upon which crops will flourish, soil drainage and structure are carefully evaluated to prevent potential challenges such as waterlogging or compaction.

Land characteristics play a crucial role in shaping the operational efficiency and viability of hi-tech agricultural ventures. The lay of the land dictates the feasibility of infrastructure installation, with considerations of topography, slope gradients, and land contours guiding decisions to optimize water management and mitigate erosion risks. Beyond the boundaries of the farm, proximity to markets and distribution channels stands as a strategic advantage. Efficient transportation logistics are vital for the timely delivery of produce to urban centers or export routes, minimizing costs and enhancing market competitiveness. Amidst the flurry of strategic considerations, environmental impact and regulatory compliance must not be overlooked. Anticipating and mitigating potential environmental impacts ensures sustainable agricultural practices, while adherence to local regulations and zoning requirements ensures operational legality and community acceptance.

Once the optimal site is identified and secured through rigorous surveying and analysis, the transformative journey of field preparation commences. From the meticulous clearing of vegetation to the precise construction of greenhouses and installation of state-of-the-art technologies, every step is imbued with a commitment to excellence. Rigorous testing and calibration of equipment ensure precise environmental control, setting the stage for the seamless integration of advanced technologies that maximize productivity and resource efficiency. In the tapestry of hi-tech agriculture, site selection and surveying stand as the cornerstone of innovation and sustainability. They pave the way for a future where precision meets passion, and where the bounty of the land meets the boundless potential of technology. Identify a suitable location based on project requirements (*e.g.*, proximity to infrastructure, environmental factors). Conduct surveys to assess terrain, soil conditions, and any potential environmental impacts.

- ☆ Determine the need for utilities such as electricity, water, and telecommunications.
- ☆ Plan for necessary connections and installations to support hi-tech equipment and operations.

Ground Preparation

In the dynamic realm of hi-tech field preparation, where innovation meets practicality, the journey begins with meticulous ground preparation. This foundational stage sets the stage for cutting-edge technologies and advanced methodologies to flourish in agriculture, construction, or any field requiring precision and efficiency. Ground preparation involves more than just clearing and leveling; it encompasses a strategic approach to optimizing soil health, infrastructure readiness, and environmental sustainability. By meticulously addressing these foundational elements, hi-tech field preparation not only lays the groundwork for immediate operational success but also ensures sustainable long-term performance, poised at the intersection of technology and practicality.

- ☆ Clear the site of any obstacles or debris that could hinder construction or operations.
- ☆ Grade the land as needed to ensure a level and stable foundation.

Construction and Installation

In the realm of hi-tech field preparation, construction and installation represent the pivotal stages where vision meets reality. This phase transcends traditional practices, integrating advanced technologies and innovative methodologies to create infrastructures that redefine efficiency and functionality. From erecting state-of-the-art facilities to installing precision-engineered equipment, every step is orchestrated with meticulous precision and a commitment to excellence. This transformative process not only builds physical structures but also lays the foundation for seamless operations, ensuring that the convergence of cutting-edge technology and strategic planning optimally supports the objectives of the project. As these structures rise from the ground, they symbolize not just progress, but the embodiment of ingenuity and forward-thinking in modern-day field preparation.

- ☆ Build facilities or structures required for the operation, such as buildings, platforms, or specialized installations.

☆ Install equipment and machinery according to technical specifications and safety standards.

Security and Safety Measures

In the realm of hi-tech field preparation, security and safety measures stand as the sentinels guarding against potential risks and ensuring the uninterrupted flow of operations. Far beyond conventional practices, these measures embody a sophisticated fusion of technology, strategic planning, and unwavering commitment to protecting valuable assets and personnel. From implementing state-of-the-art surveillance systems to designing robust access controls, every facet is meticulously crafted to fortify the integrity of the environment. This proactive approach not only safeguards against external threats but also cultivates a culture of safety that prioritizes the well-being of all involved. As advancements continue to redefine the landscape of field preparation, these measures stand as the cornerstone, ensuring resilience and continuity amidst the evolving challenges of the modern era.

☆ Implement security protocols to protect the site and its contents from unauthorized access and potential threats.

☆ Ensure compliance with safety regulations to mitigate risks associated with hi-tech equipment and operations.

Testing and Commissioning

In the intricate domain of hi-tech field preparation, testing and commissioning emerge as the critical phases that validate the integrity and readiness of sophisticated systems and infrastructure. These pivotal stages go beyond mere verification; they represent the culmination of meticulous planning and meticulous execution, aimed at ensuring that every component operates seamlessly within its intended environment. Through rigorous testing protocols and meticulous calibration, potential vulnerabilities are identified and resolved, while performance benchmarks are meticulously met or exceeded. Commissioning, in turn, marks the formal activation and integration of these systems into operational workflows, underlining a commitment to precision and reliability. As these processes unfold, they not only validate the efficacy of cutting-edge technologies but also signify the readiness to embark on a new frontier of innovation and efficiency in hi-tech field preparation.

☆ Conduct tests to verify the functionality and performance of installed systems.

☆ Commission equipment to ensure it operates as intended and meets required specifications.

Environmental Considerations

In the realm of hi-tech field preparation, environmental considerations stand as a cornerstone of responsible and sustainable practices. Beyond the pursuit of technological advancement and operational efficiency, this crucial aspect recognizes the interconnectedness between human activities and the natural world. Every decision—from site selection to operational strategies—is guided by a profound respect for ecological balance and the preservation of natural resources. Environmental considerations in hi-tech field preparation encompass a multifaceted approach. It involves meticulous assessments of potential impacts on air, water, and soil quality, alongside proactive measures to mitigate these effects. Strategies may include the integration of renewable energy sources, implementation of efficient water management systems, and adoption of practices that minimize carbon footprint and waste generation. Moreover, adherence to stringent environmental regulations and standards ensures compliance with local and international laws, fostering a harmonious relationship between technological progress and environmental stewardship. By prioritizing sustainability in every phase of field preparation, from initial planning to ongoing operations, hi-tech endeavours not only aim for operational excellence but also aspire to leave a positive legacy for future generations. Ultimately, environmental considerations in hi-tech field preparation underscore a commitment to holistic development—one that embraces innovation while safeguarding the delicate balance of our planet's ecosystems. As technologies evolve and aspirations grow, this steadfast dedication to environmental stewardship remains foundational to achieving enduring success and global sustainability goals.

- ✰ Address any environmental concerns through mitigation measures and compliance with regulatory requirements.

- ✰ Implement practices for sustainability and minimize the project's environmental footprint.

Documentation and Handover

In the intricate landscape of hi-tech field preparation, documentation and handover represent the pivotal bridge between meticulous planning and seamless execution. These critical phases are more than administrative formalities; they encapsulate the culmination of efforts to ensure clarity, accountability, and operational readiness. Documentation serves as the comprehensive repository of project blueprints, technical specifications, and operational protocols, meticulously capturing every detail from inception to completion. Handover, on the other hand, signifies the formal transfer of responsibilities and assets from project teams to operational stakeholders. It marks the culmination of rigorous testing, calibration, and validation processes, affirming that all systems are primed for optimal performance. This transition is guided by a commitment to precision, ensuring that knowledge transfer and training programs equip operational teams with the expertise to seamlessly integrate and manage hi-tech infrastructure. Together, documentation and handover not only uphold standards of transparency and compliance but also foster a culture of continuous improvement and accountability. They embody a commitment to excellence in hi-tech field preparation, where thoroughness and attention to detail pave the way for sustainable success and innovation. As technologies evolve and projects unfold, these foundational practices ensure that the legacy of efficiency and effectiveness endures beyond the initial phases, shaping a future where precision and progress converge seamlessly.

- ☆ Maintain comprehensive records of the project, including plans, specifications, and operational manuals.
- ☆ Conduct a formal handover process to transfer responsibility to operational teams or stakeholders.

Maintenance and Upkeep

In the realm of hi-tech field preparation, maintenance and upkeep emerge as vital pillars that sustain the longevity and optimal performance of advanced systems and infrastructure. Beyond the initial stages of construction and commissioning, these phases represent an ongoing commitment to reliability, efficiency, and safety. Maintenance encompasses proactive measures aimed at preventing downtime and ensuring continuous operation, while upkeep involves routine inspections, repairs, and upgrades to adapt to evolving technological advancements and operational needs. At the core of maintenance and upkeep lies a strategic approach to asset management, encompassing everything from specialized equipment to intricate networks and environmental controls. Rigorous adherence to manufacturer guidelines, coupled with predictive maintenance techniques and data-driven analytics, enables proactive identification of potential issues before they escalate. This proactive stance not only minimizes disruptions but also optimizes resource utilization, enhancing sustainability and cost-efficiency in the long term. Furthermore, training and skill development are integral components of maintenance and upkeep efforts, ensuring that personnel possess the expertise to effectively troubleshoot, maintain, and optimize hi-tech systems. By fostering a culture of continuous improvement and knowledge sharing, organizations can maximize the operational lifespan of their investments while maintaining compliance with regulatory standards and environmental responsibilities. Ultimately, maintenance and upkeep in hi-tech field preparation underscore a commitment to reliability, resilience, and operational excellence. As technologies evolve and challenges emerge, these foundational practices ensure that hi-tech infrastructure continues to deliver value, efficiency, and innovation, setting new benchmarks for sustainable development and future-proofing in the dynamic landscape of modern agriculture and beyond.

- ☆ Develop a maintenance schedule and protocols to ensure ongoing performance and longevity of equipment and infrastructure.
- ☆ Train personnel on maintenance procedures and troubleshooting techniques.

Depending on the specific application of "Hi-tech Field Preparation" These steps may vary in emphasis or detail. The goal is always to create a functional and efficient environment that supports the use of advanced technology effectively and safely.

Planting Methods

The system of planting to be adopted is selected after considering the slope of land, purpose of utilizing the orchard space, convenience *etc.* Generally, six systems of planting are recommended for fruit trees.

Square System

This system is extensively adopted and is regarded as the most straightforward among all systems. This approach involves dividing the plot into square sections and planting trees at the four corners of each square, arranged in straight rows that run perpendicular to each other. When designing the plot, a baseline is initially drawn parallel to the road, fence, or neighboring orchard. This baseline is located at a distance that is equal to half the spacing that will be used between the trees. On this line, pegs are securely attached at the specified intervals. Right angles are formed at both ends of the base line using the 3, 4, 5 meter technique commonly used by carpenters. Once three lines have been formed, it becomes simple to attach additional pegs at the desired intervals between the lines to designate the sites of the trees. This can be achieved by connecting the pegs of the lines with ropes, running in opposing directions. This technology facilitates the execution of intercultural

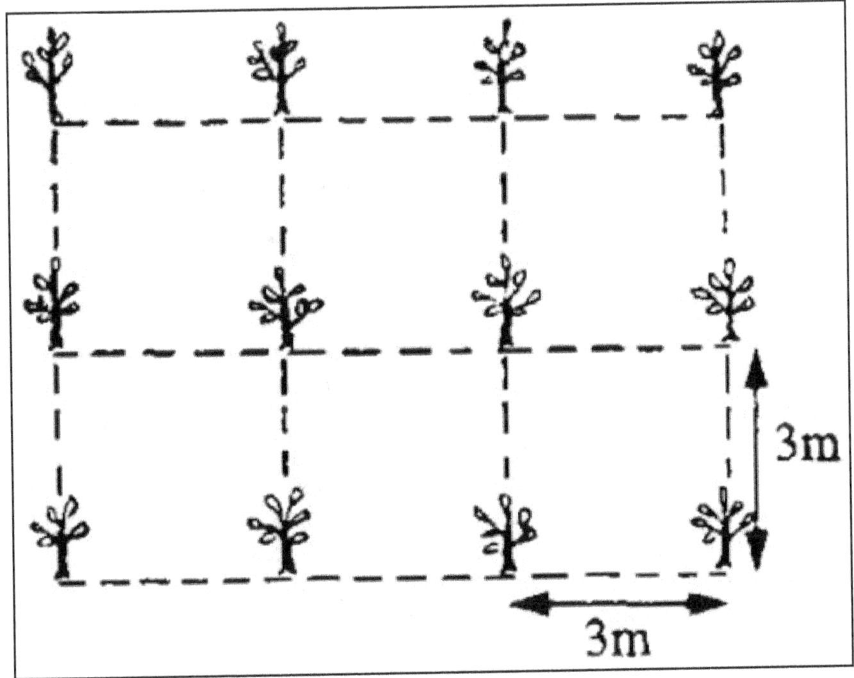

operations, including as spraying and harvesting, in a convenient and effortless manner. It is feasible to plant fast-growing fruit trees such as papaya, banana, and guava in the early stages of the orchard's development. Cultivating inter-crops such as vegetables, ginger, turmeric, cumin, coriander, and other spices can be easily accomplished. Cultivation and irrigation can be conducted in two orientations.

Rectangular System

This approach utilizes rectangular divisions instead of square divisions for the plot, with trees planted in straight rows at the four corners of each rectangle, running perpendicular to each other. Here, the same benefits that were discussed in the square system are also present. The sole distinction lies in the fact that this arrangement allows for a greater number of plants to be housed within each row, while simultaneously maintaining a wider gap between the rows.

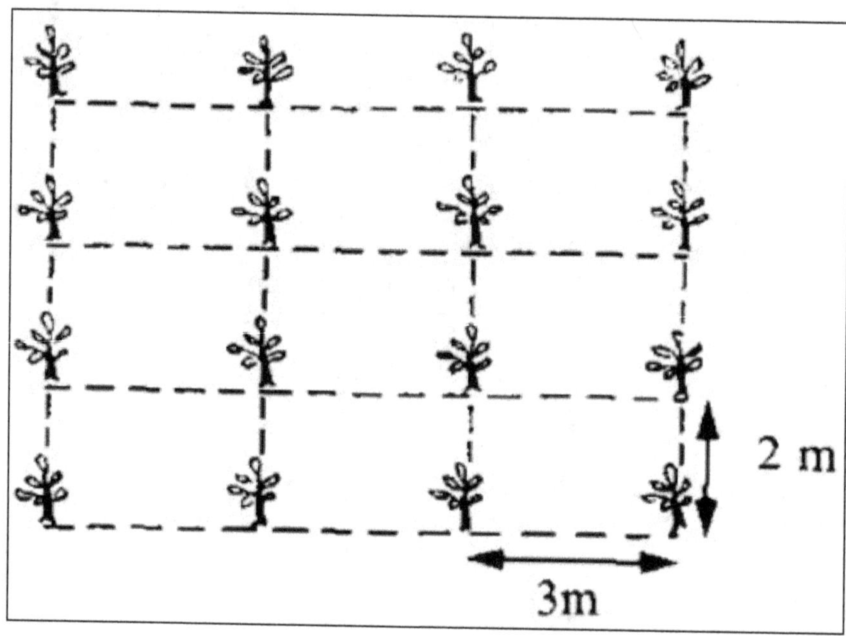

Hexagonal System

In this arrangement, the trees are positioned at the vertices of an equilateral triangle, forming a hexagon with one additional tree at the center. This approach is typically employed in areas where land is expensive and highly fertile, and where there is abundant access to irrigation water. Despite the fact that this method allows for 15 per cent more trees to be planted in a given space compared to the square system, fruit farmers typically do not choose to use it because to the challenges associated with layout and cultivation in the plot, which are not as straightforward as in the square system. To establish the plot's layout, a baseline is created on one side, similar to the square system. Subsequently, a heavy wire or chain is used to construct an equilateral triangle with rings positioned at each corner, where the length of each side matches the desired distance. Subsequently, two of these rings are positioned on the stakes of the base line, while the placement of the third ring determines the location of a tree in the second row. Subsequently, the aforementioned row serves as the foundation, and pegs are positioned inthe third row. This approach presents the complete storyline.

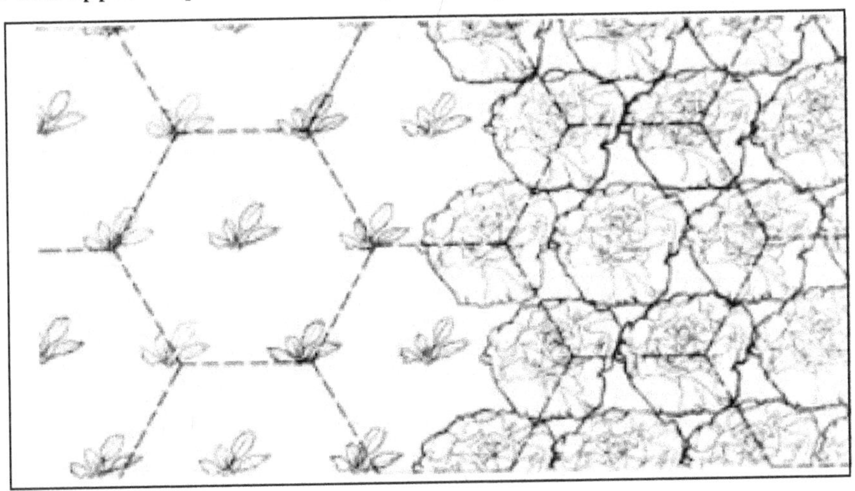

Quincunx System

This planting method for fruit trees is akin to the square system, with the addition of a fifth tree being placed at the center of each square. Consequently, the density of trees in a given area is nearly twice as high as in the square arrangement. The tree located in the center is commonly referred to as a "filler". The fillers are often fast-growing, precocious, and upright fruit trees such as banana, papaya, pomegranate, *etc.*, which are removed promptly after the major fruit trees planted at the corner of the square start producing fruit. Planting filler trees in the early stages of the orchard's development offers the grower an extra source of income.

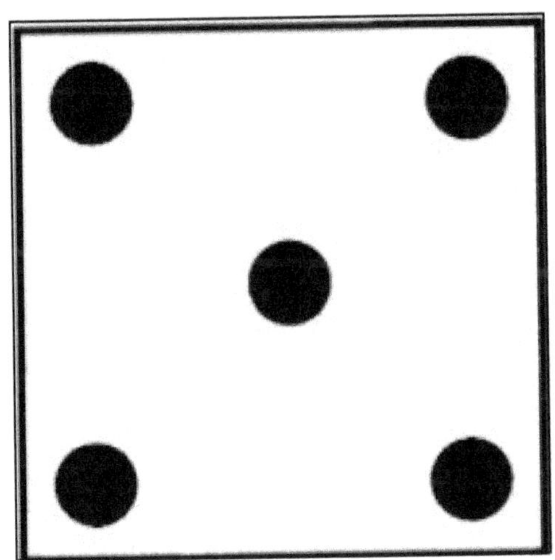

Contour System

Typically, it is observed in mountainous areas with steep inclines. It is especially well-suited for a region with hilly terrain, where there is a higher risk of erosion and where irrigating the orchard is challenging. The primary objective of this technique is to mitigate land erosion and preserve soil moisture in order to create a suitable slope for cultivating crops. The contour line is intentionally shaped and leveled to regulate the speed of water flow in the irrigation channel, allowing sufficient time for water to seep into the soil without creating erosion.

Chapter 5

Protected Cultivation

Introduction: An Overview

Protected farming refers to the practice of cultivating crops within a regulated environment. This implies that the temperature, humidity, light, and other relevant elements can be controlled according to the specific needs of the crop. This contributes to a more robust and abundant crop. Protected cultivation encompasses a range of diverse approaches. Commonly employed techniques include forced ventilation in greenhouses, natural ventilation in polyhouses, insect-proof net houses, shade net houses, plastic tunnels, and the use of mulching, raised beds, trellising, and drip watering. These measures can be employed either individually or in conjunction to create a favorable environment for protecting plants from adverse weather conditions and prolonging the duration of cultivation or off-season agricultural

Protected cultivation is a method of growing crops in a controlled environment, allowing for the regulation of factors such as temperature, humidity, and light based on the specific requirements of the crop. This controlled environment helps promote healthier plants and increases the overall yield. There are various types of protected cultivation practices, including forced ventilated greenhouses, naturally ventilated polyhouses, insect-proof net houses, shade net houses, plastic tunnels, and techniques such as mulching, raised beds, trellising, and drip irrigation. These practices can be utilized independently or in combination to create a favorable growing environment, shielding plants from harsh climates and extending the cultivation period or enabling offseason crop production. The adoption of drip irrigation in combination with raised beds and mulch films offers benefits such as weed control and improved soil moisture retention by reducing evaporation losses.

production. The implementation of drip irrigation in conjunction with raised beds (as discussed in Unit 4) and the use of mulch films serves the dual purpose of weed control and long-term moisture retention in the soil by reducing evaporation losses.

Protected or greenhouse agriculture refers to the advanced technique of cultivating crops in a controlled environment. The crops are shielded from abiotic factors such as temperature, rain, wind, humidity, and other environmental conditions, as well as biotic factors such as illnesses and insect-pests. Greenhouse production of vegetables provides growers with significant benefits in terms of superior quality, increased productivity, and favorable market prices. Greenhouse growing of vegetables during the off season can significantly boost the income of vegetable growers. This is because the vegetables grown during the regular season often do not provide high profits due to their abundant availability in the market. Greenhouse technology is widely used in countries such as the United States, Canada, and Europe. It enables the practice of precision farming and mitigates the constraints of limited space and the negative impacts of climate change. The global production of vegetables in controlled environments is on the rise. These systems regulate multiple environmental parameters, including air quality, temperature, humidity, and atmospheric gas composition. Protected cultivation, such as using greenhouses, net houses, low tunnels, mulches, *etc.*, provides numerous benefits for growing high-quality and high-yield crops, allowing for more efficient use of land and resources.

Importance of Protected Cultivation

Despite agriculture being the mainstay of India's economy for centuries, our observations over the past 50 years suggest a correlation between agricultural practices, their growth, and overall economic well-being. The trajectory of agricultural expansion indicates a combination of notable accomplishments on one side and overlooked prospects on the other. In order for India to maintain self-sufficiency and ensure food security for the impoverished population, as well as facilitate the exportation of top-notch fruits and vegetables, it is imperative to adopt novel and efficient production technologies that can consistently enhance the productivity, profitability, and reputation of the agricultural industry. An example of such an area is protected cultivation technology, which is extensively utilized in developed countries but has limited adoption in India.

The country's diversified terrain and vast range of meteorological conditions enable a multitude of agricultural patterns. India frequently encounters climate extremes, including floods, droughts, and other irregular weather patterns, which regularly lead to crop losses and economic damages for farmers. Concurrently, there has been a rise in the demand for high-quality agricultural produce in the past decade. This offers enhanced prospects for Indian farmers to embrace protected farming systems based on the specific region and appropriateness of the crops.

Commercially, greenhouses are utilized for the cultivation of non-native and off-season foods, high-quality cut flowers for export, and the propagation of superior seedlings. The economic gains derived from cultivating high-value agricultural crops can be significantly enhanced by employing greenhouse cultivation methods. To minimize the presence of chemical residues on crop produce, it is advisable to limit the use of chemical pesticides and insecticides in protected environments for crops. Greenhouses are mostly utilized as protective structures against rain, especially in parts of India with significant levels of precipitation, such as the North-eastern states and coastal areas.

Objectives of Protected Cultivation

Protected cultivation refers to the practice of growing crops within a controlled environment, often using structures such as greenhouses, shade houses, or tunnels. The objectives of protected cultivation include:

- ☆ **Extend Growing Seasons:** One of the primary goals is to extend the growing season beyond what is possible in open fields. This allows for earlier planting in the spring and later harvesting in the fall, thereby increasing overall crop yield and ensuring a more continuous supply to the market.

- ☆ **Climate Control:** Protecting crops from adverse weather conditions such as frost, wind, excessive rain, or extreme temperatures. This control helps in creating an optimal microclimate for plant growth, which can enhance yield and quality.

- ☆ **Optimize Growing Conditions:** Providing ideal conditions for plant growth by controlling factors like temperature, humidity, light intensity, and photoperiod. This enables farmers to grow crops that are otherwise sensitive to local climate variations.

- ☆ **Water Management:** Efficient use of water through irrigation systems tailored to the needs of specific crops. Protected cultivation allows for precise water application, reducing water wastage and optimizing crop water use efficiency.

- ☆ **Pest and Disease Management:** Limiting the exposure of crops to pests and diseases, which can be particularly challenging in open field conditions. Physical barriers and controlled environments can reduce the need for chemical pesticides and promote biological pest control methods.

- ☆ **Improved Crop Quality:** Enhancing the quality of produce by minimizing blemishes, improving uniformity in size and shape, and ensuring consistent flavour and texture. Controlled environments can also promote higher nutrient content in certain crops.

- ☆ **Higher Yield:** Increasing crop yield per unit area compared to open field farming, due to the ability to provide optimal growing conditions and protection from environmental stresses. This can result in higher profitability for farmers.

☆ **Diversification of Crop Production:** Allowing for the cultivation of a wider range of crops that might not otherwise thrive in a particular region or season. This diversification can reduce dependency on a single crop and provide opportunities for niche market products.

☆ **Research and Development:** Providing a platform for experimentation and innovation in agriculture, particularly in developing new varieties, techniques, and technologies that can further enhance productivity and sustainability.

☆ **Sustainable Agriculture:** Supporting sustainable agricultural practices by reducing resource use (such as water and fertilizers), minimizing environmental impacts, and promoting efficient land use.

Protected cultivation aims to optimize crop production by creating a favourable environment that maximizes yield, quality, and profitability while minimizing risks associated with climate variability and pest pressure. Currently in India, small and medium-sized farmers are engaging in flower and vegetable growing using various types of modular protected structures, based on their financial capabilities and the market conditions in their respective areas. Out of all the methods used to safeguard crops, greenhouse farming offers the greatest advantages. The primary crops cultivated in covered structures encompass floriculture crops such as rose, gerbera, carnation, anthurium, lilium, orchids, chrysanthemum, *etc.*, as well as vegetable crops including tomato, yellow and red bell peppers (from the capsicum family), cucumber, green and exotic vegetables, *etc.*

Advantages of Protected Cultivation

Protected cultivation offers significant advantages for agricultural production by providing a controlled environment that enhances crop quality, extends growing seasons, and maximizes yields. By shielding plants from adverse weather conditions such as frost, excessive rain, and wind, it creates a stable microclimate that promotes optimal growth. This controlled environment also reduces pest and disease pressures, minimizing the need for chemical pesticides and enhancing food safety. Additionally, efficient water and nutrient management within protected structures improves resource use efficiency and supports sustainable farming practices. The ability to grow a wider variety of crops, including those that are not typically suited to local climates or seasons, diversifies production and meets market demands for high-quality, off-season produce. Overall, protected cultivation enables farmers to achieve higher productivity, better crop uniformity, and increased profitability while reducing environmental impacts associated with conventional farming practices.

Protected cultivation offers several advantages for horticultural crops:

☆ **Extended Growing Seasons:** Greenhouses and similar structures provide a controlled environment that extends the growing season. This allows for earlier planting and later harvesting, resulting in multiple crop cycles per year and increased overall yield.

- ☆ **Climate Control:** Protection from adverse weather conditions such as frost, excessive rain, wind, and extreme temperatures ensures a stable microclimate conducive to optimal plant growth. This control enhances productivity and reduces crop losses due to weather-related stress.

- ☆ **Enhanced Crop Quality:** Controlled environments promote better crop quality by reducing blemishes, improving uniformity in size and shape, and maintaining consistent flavor and nutritional content. This makes produce more attractive to consumers and potentially commands higher prices in the market.

- ☆ **Water Use Efficiency:** Efficient irrigation systems within protected structures minimize water wastage by delivering water directly to the roots of plants. This reduces overall water consumption and supports sustainable agricultural practices.

- ☆ **Reduced Pest and Disease Pressure:** Physical barriers and controlled environments help prevent pests and diseases from affecting crops, reducing the need for chemical pesticides. Integrated pest management techniques can be more effectively implemented, minimizing environmental impacts.

- ☆ **Higher Yield per Unit Area:** Optimal growing conditions and reduced environmental stresses result in higher yields per square meter compared to open-field cultivation. This increases profitability for farmers by maximizing production from limited land resources.

- ☆ **Diversification of Crop Production:** Protected cultivation allows for the production of a wider range of crops that may not be suitable for local climates or seasons. This diversification can meet market demand for off-season produce and niche products, providing economic opportunities.

- ☆ **Improved Resource Use Efficiency:** Controlled environments enable precise management of inputs such as fertilizers, reducing nutrient runoff and soil degradation. This supports sustainable agriculture practices and reduces environmental impact.

- ☆ **Research and Innovation:** Greenhouses provide a platform for research and development of new crop varieties, growing techniques, and technologies. Innovations in protected cultivation contribute to continuous improvements in agricultural productivity and sustainability.

- ☆ **Market Advantage:** High-quality, consistent produce from protected cultivation often commands premium prices in markets where consumers prioritize quality and freshness. This can enhance market competitiveness and profitability for growers.

Protected cultivation in horticultural crops offers numerous advantages by creating optimal growing conditions, enhancing crop quality and yield, conserving resources, and supporting sustainable agricultural practices. These benefits contribute to the economic viability and resilience of agricultural production systems in diverse climates and regions.

Limitations of Protected Cultivation

Protected cultivation, while advantageous for enhancing crop quality and extending growing seasons, presents several challenges. The initial setup costs can be prohibitive, requiring significant investments in infrastructure and technology. Ongoing operational expenses, including energy, water, and labor, can also be high, affecting economic viability, particularly for less profitable crops. Technical expertise is essential for managing climate control systems and preventing pest outbreaks, adding to the complexity. Moreover, despite efforts to control the environment, fluctuations in temperature and humidity within greenhouses or tunnels remain difficult to completely eliminate. Environmental concerns, such as increased energy consumption and potential waste generation, further underscore the need for sustainable practices. Regulatory hurdles and market competition from traditional farming methods also pose challenges, influencing the overall feasibility and adoption of protected cultivation methods. While protected cultivation offers significant advantages, it also has several limitations and challenges:

- ☆ **Initial Investment Costs:** Setting up a protected cultivation system can require a substantial upfront investment in infrastructure such as greenhouse structures, irrigation systems, heating or cooling systems, and automation technology. This initial cost can be a barrier for small-scale farmers or those in developing regions.

- ☆ **Operating Costs:** Maintaining controlled environments involves ongoing expenses for energy (heating, cooling, lighting), water, nutrients, and labor. These operational costs can be higher compared to traditional open-field farming, impacting the economic feasibility, especially for crops with low market value.

- ☆ **Technical Expertise:** Successful management of protected cultivation systems requires specialized knowledge and skills in crop management, climate control, pest and disease management, and greenhouse operation. Farmers may need training and ongoing support to effectively manage these systems.

- ☆ **Climate Control Challenges:** Despite efforts to control the environment, challenges such as fluctuations in temperature, humidity, and light intensity can still occur within greenhouse or tunnel environments. Maintaining optimal conditions throughout the year can be difficult, especially in regions with extreme weather conditions.

- ☆ **Pest and Disease Management:** While protected cultivation can reduce some pest and disease pressures, it can also create a favorable environment for certain pests and diseases that thrive in closed systems. Integrated pest management strategies must be carefully implemented to prevent outbreaks and minimize chemical inputs.

☆ **Dependency on Technology:** The reliability of technology such as climate control systems, irrigation systems, and sensors is crucial for the success of protected cultivation. Technical failures or malfunctions can have significant consequences for crop health and productivity.

☆ **Limited Crop Diversity:** Not all crops are suitable for protected cultivation, and some crops may not achieve the same quality or yield as in open-field conditions. This limitation restricts the range of crops that can be economically grown using these methods.

☆ **Environmental Concerns:** The use of energy-intensive heating, cooling, and lighting systems in protected cultivation can have environmental impacts, including increased carbon footprint and energy consumption. Efforts to improve energy efficiency and sustainability are ongoing but remain a concern.

☆ **Regulatory and Permitting Issues:** Depending on the region, there may be regulatory requirements and permitting processes for establishing and operating protected cultivation facilities. Compliance with environmental regulations and zoning laws can add complexity and cost to the operation.

☆ **Market Access and Competition:** Despite potential advantages in quality and yield, producers engaged in protected cultivation may face competition from traditional open-field producers. Market acceptance and differentiation of protected cultivation products may vary depending on consumer preferences and market trends.

Advantages of Greenhouse

☆ Greenhouses represent a cornerstone of modern agricultural and horticultural practices, offering a controlled environment that confers several significant advantages. Firstly, they extend the growing season by shielding plants from adverse weather conditions such as frost, wind, excessive rain, and hail, thereby allowing for earlier planting and later harvesting. This extension not only increases the overall yield per year but also ensures a more consistent and reliable supply of crops, regardless of external climate fluctuations.

☆ Climate control within greenhouses is another crucial advantage. Farmers can regulate temperature, humidity, light intensity, and ventilation to create an optimal microclimate for plant growth. This precision enables cultivation of crops that might otherwise struggle in the local climate, leading to improved productivity and quality. By maintaining stable conditions, greenhouses promote uniform growth, enhance color development, and increase the nutritional content of crops, which are all factors that contribute to higher market value and consumer appeal.

☆ Water management in greenhouses is highly efficient compared to traditional open-field methods. Systems such as drip irrigation, hydroponics, or automated watering systems allow for precise delivery of water and nutrients directly to the roots of plants. This targeted approach minimizes water wastage, reduces runoff, and optimizes nutrient uptake, thereby supporting sustainable agriculture practices and conserving natural resources.

☆ Greenhouses provide robust protection against pests, diseases, and extreme weather events. The physical barrier they create limits the entry of pests and reduces the spread of diseases, reducing the need for chemical pesticides and promoting integrated pest management (IPM) strategies. This environmental benefit not only safeguards crop health but also mitigates potential risks to human health and ecosystems associated with pesticide use.

☆ The versatility of greenhouses is another advantage, as they enable the cultivation of a wide range of crops throughout the year. This includes exotic or specialty crops that may not be viable in the local climate or season, thus diversifying agricultural production and meeting diverse consumer demands. Additionally, the intensive use of space within greenhouses allows for higher yields per unit area compared to open fields, maximizing land productivity and economic returns for farmers.

☆ Research and innovation thrive within greenhouse environments, where controlled conditions facilitate experimentation with new crop varieties, cultivation techniques, and technologies. This continuous improvement fosters advancements in agricultural productivity, sustainability, and resilience to climate change, supporting the long-term viability of farming operations.

☆ Greenhouses offer a suite of advantages that contribute to enhanced crop productivity, improved quality, resource efficiency, and profitability in agriculture and horticulture. They play a crucial role in meeting global food demand while minimizing environmental impact, making them indispensable tools in modern farming practices.

Various Schemes for Protected Cultivation

Protected horticultural cultivation is supported by various schemes and initiatives aimed at promoting sustainable agriculture, enhancing productivity, and ensuring food security. Some of the notable schemes and programs include:

☆ **National Horticulture Mission (NHM):** Launched by the Government of India, NHM aims to promote holistic growth of the horticulture sector by providing support for creating infrastructure, enhancing production capabilities, improving post-harvest management, and promoting integrated pest management practices.

☆ **Pradhan Mantri Krishi Sinchayee Yojana (PMKSY):** This scheme focuses on improving water use efficiency in agriculture, including horticulture. It promotes the adoption of micro-irrigation systems such as drip and sprinkler irrigation, which are beneficial for water conservation and efficient nutrient delivery in protected cultivation.

- ☆ **Mission for Integrated Development of Horticulture (MIDH):** MIDH is another centrally sponsored scheme aimed at promoting holistic growth of the horticulture sector through area-based and demand-driven approach. It supports the development of infrastructure for protected cultivation, creation of water resources, and adoption of advanced technologies.

- ☆ **Greenhouse Scheme (Under MIDH):** Specifically under MIDH, there are provisions for financial assistance to farmers for setting up greenhouses and polyhouses. This includes subsidies on greenhouse structure, equipment, and inputs for protected cultivation of high-value horticultural crops.

- ☆ **Rashtriya Krishi Vikas Yojana (RKVY):** RKVY supports states in increasing agricultural productivity and ensuring food security through various interventions, including infrastructure development for protected cultivation, training of farmers, and adoption of best practices in horticulture.

- ☆ **Paramparagat Krishi Vikas Yojana (PKVY):** Under this scheme, organic farming practices are promoted, which can include organic protected cultivation techniques. Farmers are encouraged to adopt organic inputs and methods to reduce chemical usage and promote sustainable agricultural practices.

- ☆ **Technology Mission for Integrated Development of Horticulture in North Eastern States, Sikkim, and Jammu and Kashmir:** This mission targets specific regions to enhance horticultural production through protected cultivation, organic farming, and infrastructure development tailored to local climatic conditions.

- ☆ **Subsidy Schemes for Horticulture in State Governments:** Many state governments in India and other countries offer their own subsidy schemes and incentives for promoting protected horticultural cultivation. These schemes often include financial assistance for infrastructure development, equipment purchase, and training programs.

- ☆ **Research and Development Initiatives:** Government and private sector investments in research and development play a crucial role in advancing technologies and innovations for protected cultivation. These initiatives focus on improving crop varieties, optimizing climate control systems, and enhancing sustainable farming practices.

- ☆ **International Funding and Collaborations:** Global organizations and foreign governments often collaborate with local authorities to support protected horticultural cultivation in developing countries. This can include funding for infrastructure, technology transfer, and capacity building programs.

These schemes and initiatives collectively aim to create an enabling environment for farmers to adopt protected cultivation techniques, improve productivity, ensure food security, and promote sustainable agricultural practices in horticulture.

Sustainable Horticultural Crops for Protected Cultivation

Protected farming is suited for a diverse array of horticulture crops. Commonly cultivated crops in protected settings include:

- ☆ **Tomatoes:** Tomatoes are highly favored for cultivation under protected conditions. Greenhouses offer ideal conditions for plant growth, resulting in increased yields and superior fruit quality when compared to cultivation in open fields.

- ☆ **Cucumbers:** Cucumbers flourish in sheltered conditions, particularly in greenhouses or high tunnels. Optimal climate control and effective pest and disease management enhance crop productivity and enhance the quality of fruits.

- ☆ **Peppers:** Both sweet and hot peppers are suitable for cultivation in protected environments. Pepper plants thrive in greenhouses due to the optimal circumstances they provide, which result in longer growth seasons and higher yields.

- ☆ **Leafy greens:** Leafy greens such as lettuce, spinach, kale, and arugula can be cultivated throughout the year in sheltered areas. Greenhouses offer insulation against high temperatures and enable meticulous regulation of moisture levels.

- ☆ **Various fruits:** Various fruits such as grapes, apples, pears, peaches, plums, cherries, and strawberries are typically cultivated in greenhouses or high tunnels. These structures offer shelter from weather, pests, and illnesses. This enables enhanced fruit quality and prolonged harvest seasons.

- ☆ **Flowers:** Numerous floral species, including roses, chrysanthemums, and gerberas, are well-suited for cultivation in protected environments. Greenhouses provide a controlled environment and shelter from adverse weather conditions, leading to the production of superior flowers. Examples of flowers include rose, carnation, gerbera, anthurium, lilium, orchids, and chrysanthemum.

- ☆ **Certain herbs:** Certain herbs, such as basil, cilantro, mint, and parsley, can be easily grown in sheltered environments. Controlled settings guarantee uniform growth, increased yields, and superior quality herbs.

☆ **Melons:** Specific cultivars of melons, such as muskmelons and cantaloupes, can thrive when cultivated in sheltered locations. Greenhouses offer precise regulation of temperature and humidity, resulting in enhanced fruit quality.

☆ **Beans:** Certain varieties of legumes, such as green beans and runner beans, have the potential to be cultivated within enclosed structures for protection. These habitats provide shelter from unfavorable weather conditions and pests, leading to enhanced yield. The suitability of certain crops for protected cultivation might vary based on factors such as area climate, market demand, and available resources. When choosing crops for protected cultivation, it is important to take into account the preferences of farmers and the specific requirements of the local area.

Future Prospects of Protected Cultivation of Horticultural Crops

Protected cultivation is expected to play an increasingly important role in meeting the global demand for fresh, high-quality horticultural produce in the coming years. This is due to a number of factors, including:

☆ The growing population and urbanization

☆ The increasing demand for healthy and nutritious food

☆ The need to produce food in a sustainable manner

Factors Affecting the Adoption of Protected Cultivation of Horticultural Crops

Multiple variables impact the adoption of protected growing of horticultural crops. The following items are included:

☆ **Expenses:** The initial capital and continuous expenses related to establishing and upkeeping shielded cultivation structures might be substantial. This encompasses costs associated with infrastructure, equipment, materials, and specialist technologies. The cost-effectiveness of these expenditures can influence the farmers' inclination to embrace protected farming techniques.

☆ **Knowledge and Skills:** The adoption of protected cultivation techniques necessitates specialized knowledge and skills that may differ from those used in traditional open-field farming methods. It is essential for farmers to comprehend concepts such as temperature control, irrigation systems, pest management, and crop nutrition in a controlled setting. Inadequate knowledge and skills might impede the process of adoption.

☆ The profitability and market demand for horticultural crops cultivated utilizing protected growing techniques are important factors in determining their acceptance. Farmers require certainty over the ability to sell their agricultural products at advantageous prices, which will compensate for the extra expenses linked to protected farming. Conducting market research and analysis is crucial for assessing the feasibility and level of demand for these crops.

☆ The appropriateness of the local climatic and environmental conditions for protected farming is an additional aspect that affects the acceptance of this practice. Protected structures offer insulation from severe weather conditions, but they necessitate sufficient sunlight, water supply, and appropriate temperature ranges. Evaluating whether the local conditions are in line with the prerequisites of protected cultivation is crucial.

☆ Access to resources is essential for the effective adoption of a project. The availability and accessibility of resources, including land, water, energy, and trained labor, play a key role in this regard. Sufficient land area, a dependable water supply, and availability to electricity or alternative energy sources are essential prerequisites for establishing and running protected agricultural systems. Having a sufficient supply of skilled workers or the capability to provide training to employees is also crucial.

☆ **Government Support and Policies:** Favorable government policies, financial aid, incentives, or technical guidance can greatly promote the adoption of protected horticulture. These programs assist farmers in handling the upfront expenses, provide training and venues for sharing knowledge, and establishing a favorable atmosphere for adopting protected agricultural methods.

☆ **Risk Management:** Farmers may exhibit caution in embracing protected cultivation due to perceived risks and uncertainty. Profitability can be affected by factors such as crop failure, disease outbreaks, insect infestations, or market volatility. Implementing risk management techniques, such as crop insurance, technical assistance, and information accessibility, can effectively reduce these concerns and promote the adoption of such strategies. The widespread implementation of protected cultivation for horticulture crops is contingent upon a confluence of economic, technological, commercial, and environmental factors. To promote wider adoption of protected cultivation methods, it is important to address these factors through education, financial support, market growth, and risk management measures.

Chapter 6

Micro-Irrigation Systems and its Components

Introduction: An Overview

Micro-irrigation systems represent a transformative advancement in agricultural and landscaping practices, offering efficient water management solutions that significantly enhance crop yield, conserve water resources, and mitigate environmental impact. At its core, a micro-irrigation system operates on the principle of delivering precise amounts of water directly to the root zone of plants, thereby maximizing water utilization while minimizing wastage. This innovative approach stands in stark contrast to traditional irrigation methods, which often result in excessive water loss due to evaporation, runoff, and inefficient distribution. Comprising a sophisticated network of components meticulously designed to cater to diverse agricultural and horticultural needs, a micro-irrigation system orchestrates a symphony of technology, engineering, and agronomic expertise. Its components seamlessly integrate to create a holistic solution tailored to the specific requirements of crops, soil types, and environmental conditions. From the intricate network of pipes and tubing to the intricate emitters and valves, each element plays a crucial role in optimizing water delivery, fostering root development, and nurturing plant growth.

Central to the functionality of a micro-irrigation system are its various components, which encompass a diverse array of elements ranging from emitters and filters to controllers and sensors. Emitters, the frontline soldiers of water distribution, regulate the flow of water with precision, ensuring that each plant receives its requisite share. Filters act as guardians, purifying the water supply and safeguarding the system against clogging and damage. Meanwhile, controllers and sensors serve as the brain and nervous system, respectively, orchestrating the operation of the system with unparalleled efficiency and responsiveness. Furthermore, the adaptability and scalability of micro-irrigation systems render them indispensable across a spectrum of agricultural contexts, from small-scale subsistence farming to large-scale commercial enterprises. Their modular design allows for customization and expansion, empowering farmers and landscapers to tailor solutions that align with their evolving needs and resource constraints. Moreover, by minimizing water usage and maximizing crop productivity, micro-irrigation systems not only bolster agricultural resilience in the face of climatic uncertainties but also foster sustainability by reducing the ecological footprint of irrigation practices. The advent of micro-irrigation systems heralds a paradigm shift in the realm of water management, offering a beacon of hope for a world grappling with the dual challenges of food security and water scarcity. As humanity grapples with the imperative to feed a burgeoning population amidst dwindling water resources and escalating environmental pressures, the adoption of micro-irrigation systems emerges as a beacon of sustainable progress, embodying the harmonious integration of technology, ecology, and agricultural ingenuity.

Micro-irrigation systems can be configured in several manners. In order for a system to irrigate effectively, the distribution of water must be consistent. The discharge levels of the emitters with the lowest and maximum output should not differ by more than 10 percent. In order to accomplish this, it is necessary to ensure that the pipes and tubing are appropriately proportioned. Water at the source is managed using automatic valves, occasionally supplemented with fertilizers or chemicals, and filtered and adjusted to appropriate levels for the emitters. Subsequently, water is distributed to individual emitters via a system of polyvinyl chloride (PVC) and polyethylene (PE) pipes. The constituents of a micro-irrigation system can be classified into the subsequent overarching categories:the pumping station;

- ☆ Control systems;
- ☆ Filtration systems:
- ☆ Mainlines, submains, manifolds, and laterals;
- ☆ Flow control devices;
- ☆ Fertigation-chemigation systems; and
- ☆ Emitters.

Pumping Stations

Several variables must be taken into account prior to selecting a suitable groundwater pumping system. The feasibility of implementing a groundwater pumping system in greenhouses ultimately hinges on its financial sustainability. The cost of installing a pumping system in greenhouses can vary greatly based on factors such as the system's intended usage, hydrological conditions, groundwater quality, proximity to electrical supply, and alternatives for waste disposal. The cost of establishing a bore is influenced by factors such as the design, materials, and building process. These factors also affect the amount and quality of water obtained.

Types of Groundwater Pumps

Various groundwater pumps are utilized in greenhouse operations, each with distinct configurations that may restrict their applicability in specific circumstances.

- ☆ **Pump Curves:** Centrifugal pumps are often used for this purpose. Centrifugal pumps operate over a wide range of operating conditions but are limited by the suction lift (theoretically, 33 ft (10 m) at sea level, but in practice, about 23 ft (7 m)), and they need to be primed. Graphical characteristic curves define the operation of these pumps in terms of the discharge, the head, the size of the impeller, and the horsepower.

- ☆ **Variable Frequency Drives (VFDs):** Variable Frequency Drives (VFD) enable irrigation pump systems to work only at the rate necessary in order to complete the job for which they are responsible. Installing a VFD on a system reduces the rotational speed of the motor, which then decreases the speed of the pump, allowing it to consume exactly the amount of power required for use.

Power Units for Pumping

Power units used for irrigation pumping include diesel engines, LP gas and gasoline engines, and electric motors. Power units should be selected to match the power requirements of the pumping application. Overloading a power unit may shorten its useful life significantly, while power units too big for the job operate at reduced efficiency. The efficiency of electric motors ranges from 85 to 92 percent. Large electric motors (above 15 to 20 horsepower) are more efficient than small electric motors. Gasoline engines operate at efficiencies of 20 to 30 percent. Diesel industrial engines of the type used for irrigation operate at efficiencies of 25 to 50 percent, depending on age and condition.

- ☆ **Electric Motors:** Many irrigation-pumping plants use electric motors. They have a long-expected life, require minimal maintenance, and are very reliable. An electric motor, properly selected and protected, can be expected to supply many years of trouble-free power if properly designed and operated, including correct mounting, rodent protection, good ventilation, adequate shelter from the elements, and safety devices against overloading, under-voltage, and excessive heating. Advantages of the electric power are relatively long life of the motor, low maintenance costs, dependability, and ease of operation. Normally, an electric motor can be expected to last 5 to 10 years in continuous operation.

- ☆ **Internal Combustion Engines:** There are several factors in selecting an internal combustion engine for a pumping application, including portable or stationary, air cooled or water cooled, fuel type, speed, size, efficiency, emissions, maintenance costs, expected life, and initial cost. Liquid-cooled engines can be cooled by either a radiator or a heat exchanger. The heat exchanger can recover up to 8 per cent more horsepower than the radiator because it does not require the engine to run a fan to pull air through the radiator. The heat exchanger needs to be sized to the engine for adequate cooling even on the hottest days, but not so large that the engine overcools and forms excessive internal sludge. The manufacturer's recommendation on engine temperature should be followed closely to prevent problems from overheating or under-heating, and a properly rated coolant thermostat be used to regulate coolant temperature. Internal combustion engines can be designed to run on one or several types of fuels. Fuel chosen will probably be determined by availability and cost at any one location. Liquid petroleum gas (LPG) and natural gas are very similar fuels except that natural gas has slightly lower energy content.

Filtration Systems

Filtration equipment ensures that organic and inorganic particles such as sand, algae, or silt bigger than the smallest inlet or outlet hole in the system are removed, protecting the irrigation system from clogging. If chemicals or acids must be added for water treatment, be sure to determine if the materials should be injected before or after any filtration equipment. Some chemicals well react with the materials from which the filters are manufactured. Water soluble fertilizers are typically injected after the filters and must be in solution, and chemicals must be tested prior to injection into the irrigation system. Sometimes a combination of more than one type of filter will need to be used. Two sets of filters are often recommended. The primary filter removes suspended material from the water. A smaller-capacity secondary filter installed downstream from the primary filter protects the irrigation system if the primary filter fails. Filtration capacity is expressed in "mesh" (mesh numbers correspond to openings per inch, *e.g.*, 200 mesh has 200 openings per inch) or microns. Most micro-irrigation applications require mesh sizes between 100 and 200. Filters are solely used to remove particles from water; they cannot take out dissolved solids, salts, and other toxic elements. Check with the manufacture of the emission device for their recommendations on the mesh or micron size of filter.

Main, Submain, Manifolds and Laterals

The main objective of a micro-irrigation system is to provide an irrigation system such that when properly managed, each plant will receive the same amount of water and nutrients, in sufficient quantity, at the proper time, and as economically as possible. For this goal to be realized, the system must deliver the needed pressurized amount of water to each emitter. If the headworks and other previously described components are performing properly, mains, submains, and manifolds must then deliver the water to the laterals and emitters.

☆ **Main and Submain Lines:** The main and submain lines carry water from the control head to the manifold or directly to the lateral lines. The basic system subunit includes the manifold with attached laterals. The main line to the greenhouse is buried under ground at a safe depth below the frost line. The valve assembly that makes the transition from the main lines to submains at each block location may be above ground or in a box below ground level. Pressure control or adjustment points are provided at the inlets to the manifold. Because of these pressure-control-point locations, pipe size selection for the main and submain lines is not affected by the pressure variation allowed for the subunit.

☆ **Manifolds:** The manifold, or header, connects the mainline to the laterals. Manifold pipelines are frequently installed underground with laterals on the surface, although laterals also may be underground. Underground pipelines last longer and do not interfere with greenhouse operations. Main and manifold pipelines are often set 18 to 24 inches below the surface, but site conditions may require other depths.

☆ **Laterals:** Laterals or emitter lines supply water to the emission devices from the manifolds. Black polyethylene (PE) is generally used as the laterals and ranges in size from ½ to 1 inch (1.25 to 2.54 cm), but the vast majority of greenhouse irrigation systems use ½ inch tubing. The PE material is used because of its high strength and impact resistance properties.

Flow Control Devices

Any device installed in a fluid supply system, in order to ensure that the fluid reaches the desired destination, at the proper time, in the required amount (the flow rate), and under the right pressure, is called a control device. As such an appliance controls proper operation of a fluid system, selecting its type, size and placement is of uppermost importance and ought to be done with the full knowledge of the various features of the device and with complete understanding of the way it performs.

☆ **Valves:** Valves play key roles in controlling pressure, flow and distribution under different conditions to optimize performance, facilitate management, and reduce maintenance requirements in micro-irrigation systems. Various types of valves are used in micro-irrigation systems to protect and control the irrigation system: air and vacuum relief, flow control, pressure regulation, pressure sustaining, and safety. Valves come in various design, sizes, materials, and configuration, manual or automatic, metal or plastic, and hydraulically or electronically controlled. Valves should be chosen based on performance factors such as friction loss, maintenance, accuracy, reliability, durability, speed of closing/opening, flow range, pressure reduction ratio, simplicity, and cost.

Water Flow Meters

An important device for measuring water movement between the water source and the greenhouse is the water flow meter. Close monitoring and accurate recordkeeping with this device will allow the irrigator to make fundamental adjustments and detect problems before they can have serious effects on the plants. Flowmeters can either be monitored manually or automatically by computerized monitoring and control systems. A key requirement of operating a micro-irrigation system is knowing how much water is being supplied to the crops in the greenhouse. In-line flowmeters may register total flow in standard volumetric units such as gallons, cubic feet, acre-feet, or others.

Types of Flow Meters

There are several types of flowmeters. The most commonly used flow meter is the propeller meter. These meters are designed for specific pipe sizes and work best within a range of flow rates. It requires installation in a straight section of pipe and for the pipe to flow at full capacity to register accurately. The speed of the propeller is a function of the flow rate. The pipe must be full to obtain accurate measurements. Variation in water pressure will also alter the amount of water registered on the meter. Magnetic flow meters or "mag meters" measure flow by creating a magnetic field that senses flow and produces a signal related to the flow of water.

Pressure Gauges

The performance of micro-irrigation systems depends on consistent control and knowledge of water pressure. Regardless of how well the micro-irrigation system is designed or how well the emitters are manufactured, operating pressures must remain at design specifications to maintain the desired performance and distribution uniformity. Manually monitoring pressures often or continuously with automation is important because changes in pressure can indicate a variety of problems.

Fertigation/Chemigation Components

Most greenhouse operations are going to want to have their micro-irrigation system set up for fertigation or chemigation (applying fertilizers, pesticides, or herbicides through the irrigation system). Chemigation requires a large tank to hold the chemicals that are to be injected (usually in liquid form), a pump and an injection port that mixes the chemicals with the water. Several protections are also required to protect the water source, and the environment in case of spills or problems.

Emitters

The actual application of water in a micro-irrigation system is through an emitter that controls the flow of water from the lateral line into the substrate. The emitter decreases the pressure (reduces the head) from the lateral line to the substrate. This may be done by small holes, long passageways, vortex chambers, or other mechanical means. The quantity of water delivered from these emitters is usually expressed in gallons per hour (gph). Emitters can be divided into two categories: line-source emitters and point-source emitters.

Line-Source Emitters

Line source emitters consist of drip tubing with supply orifices to meter water before it enters the line; then, the water passes through a labyrinth of flow paths to dissipate or compensate pressure and exits to one or more distribution orifices. Line-source emitters are suitable for closely spaced row crops, with the rows separated several feet apart, as with most vegetable crops. Line source emitters use two main tubing configurations:

- ☆ **Thin-Walled Drip Line:** A thin-walled drip line, also known as drip tape, is a thin-walled polyethylene product, collapses when not pressurized, and has emitters formed into its seam during manufacturing. The thin-walled drip tape inflates upon pressurization. Drip tape systems can be applied to crops in beds where plants are growing at a specified spacing. Often drip tape is used with high value crops, such as strawberries, vegetables, flowers, or nursery stock. Drip tape can in some cases be used for more than one year in the greenhouse but is typically removed and discarded at the end of each growing season.

- ☆ **Thick-Walled Drip Hose:** The thick-walled drip hose is a robust variation of the thin walled drip line. The internal emitters are molded or glued to the drip hose. It is more durable because of its considerable thickness. The diameter of the drip hose is like that of the thin walled drip line.

Point-Source Emitters

Point-source emitters are used when widely spaced point sources of water are needed, as in the case of greenhouse crops where they are spaced several feet apart. Point-source type emitters are attached to the lateral pipe. The installer can select the desired location to suit the planting configuration or place them at equally spaced intervals.

Pressure Compensating Emitters

Pressure compensating emitters, also referred to as "PC" emitters, deliver a precise amount of water regardless of changes in pressure due to long rows or changes in terrain. The main feature of these drip emitters is the ability to continuously adjust to varying water pressures, ensuring a constant flow rate from each emitter, regardless of water pressure fluctuations or varying elevations along the line.

Non-Pressure Compensating Emitters

Non-pressure compensating emitters use a turbulent flow action which provides greater durability and longevity along with clogging resistance and low maintenance. Non-pressure compensating emitters will have varying output flow at varying inlet pressures. Therefore, the flow will vary along uneven terrain, and each dripper will emit a different amount of water depending on its location on the supply line.

Micro-Irrigation for Greenhouse Crops

Water Quality for Micro-Irrigation Systems

Water quality can have a negative effect on the performance of a micro-irrigation system due to plugging of emitters. Problems can be caused by inorganic solids (*e.g.*, silt and sand), organic solids (*e.g.*, algae, bacteria, slime) and dissolved solids (*e.g.*, calcium, iron, manganese). Potential problems can be minimized by testing the water, so you are aware of potential issues.

Collecting a Water Sample

As general guidance, when collecting a sample from a well, first allow the pump to run for at least 15 minutes (even longer is preferable). A quart sample volume is usually ample for chemical constituent analyses. Make sure the sample bottle is clean and rinsed. Collect the sample as close to the well as possible but do not collect samples too close to a chemical injection point since there may be insufficient mixing. Thoroughly flush all pipes and components prior to sampling. Surface water is more difficult to sample, and its test results can change significantly over the season.

Water Quality Analysis

A water quality analysis can give the grower a "heads up" on potential trouble areas for the micro-irrigation system. This test should be accomplished before the final design ofthe system to ensure that proper components are installedto address any problem areas. The analysis should include testing for pH, dissolved solids, manganese, iron, hydrogensulfide, carbonate and bicarbonates. The quantity and sizeof particulate matter should also be known as this will determine the aperture size of any screen filters.

Suspended Solids

Suspended solids in the water supply include soil particles ranging in size from coarse sands to fine clays, living organisms including algae and bacteria, and a wide variety of miscellaneous waterborne matter. Suspended solids can serve as carriers for organic compounds.

Chemical Precipitation

Chemical plugging usually results from precipitation of one or more of the following minerals: calcium, magnesium, iron, or manganese. The minerals precipitate from solution and form encrustations (scale) that may partially or completely block the flow of water through the dripper. Water containing significant amounts of these minerals and having a pH greater than seven has the potential to plug drippers. Particularly common is the precipitation of calcium carbonates, which is temperature and pH dependent.

Biological Growth

A micro-irrigation system can provide a favorable environment for bacterial growth, resulting in slime buildup. This slime can combine with mineral particles in the water and form aggregates large enough to plug drippers.

Maintenance of Micro-Irrigation Systems

The major cause of failure in micro-irrigation systems is emitter plugging. Emitter plugging can severely degrade irrigation system performance and application uniformity. Because the emitters are small and can easily plug, it is important to understand the maintenance requirements of these systems and be proactive to prevent plugging.

Filter Maintenance

Filters are the first line of protection for your micro-irrigation system, and they need regular maintenance to operate at a high level. A pressure drop across the filter is an indication of debris, so it must be monitored to maintain full flow. Pressure drops can be detected by installing gauges before and after the filter or a differential pressure gauge that compares the two sides. Monitor the filter pressure differential frequently, especially as water conditions change through the season.

Flushing the Main Lines, Submains, and Laterals

The main lines, submains, and particularly the lateral lines should be flushed periodically to ensure sediments are cleared from the system, which can cause a potential hazard by clogging emitters. Flushing the main, submain, and laterals will considerably reduce the accumulation of organic and mineral materials in the system. This will prevent those materials from reaching the drippers and eventually clogging them, thus minimizing the quantity of chemical products required to maintain the system. Regular flushing of the main, submains, and laterals will result in a significant saving of labor time and chemicals.

Emitter Inspection

Systematic checking is required to spot malfunctioning emitters or to use accurate flow and pressure measurements and analyze their rates of change over time. Slow clogging causing partial blockage results from sediments, precipitates, organic deposits, or mixtures of these. Physical deterioration of parts is a concern with pressure compensating emitters. The flow passage may slowly close as the compensating part wears out. Mechanical malfunction can also be a problem in flushing emitters.

Chemical Injection for System Maintenance

Unfortunately, filtration alone is not always adequate to solve all water quality problems. Chemicals are necessary to control algae, iron, and sulfur bacteria and disease organisms. Chemicals can cause some materials to settle out or precipitate out of the water, while causing other materials to maintain solubility or stay dissolved in the water. Chlorine is a primary chemical used to kill microbial activity; to decompose organic materials; and to oxidize soluble minerals, which causes them to precipitate out of solution.

EC/pH Based Irrigation/Fertigation Scheduling

EC (Electrical Conductivity) and pH-based irrigation/fertigation scheduling is a method used in agriculture to manage water and nutrient application based on the electrical conductivity and pH levels of the soil or the nutrient solution. Here's a detailed description of how this system works:

Basics of EC and pH

Electrical Conductivity (EC)

- ☆ EC measures the ability of a solution to conduct electricity, which is directly related to the concentration of dissolved salts (total dissolved solids or TDS) in the water or soil solution.
- ☆ Higher EC values indicate higher concentrations of salts and nutrients.

pH

- ☆ pH measures the acidity or alkalinity of a solution.
- ☆ Different plants have different pH preferences for optimal nutrient uptake and growth.

Principles of EC and pH-based Irrigation/Fertigation

Monitoring and Measurement

- ☆ **Continuous Monitoring:** Sensors are placed in the soil or in the nutrient solution to continuously monitor EC and pH levels.
- ☆ **Periodic Testing:** Regular sampling and testing of soil or nutrient solution are also done to verify sensor readings and ensure accuracy.

Setting Target Ranges

- ☆ **EC Range:** Based on the crop's nutrient requirements, optimal EC ranges are established. These ranges ensure that plants receive adequate nutrients without risking salt accumulation.

☆ **pH Range:** Similarly, pH ranges are set to ensure optimal nutrient availability for the specific crop being grown.

Decision Making

☆ **Fertigation Adjustments:** Based on real-time or recent measurements, adjustments are made to the nutrient solution composition. If EC is too low, more nutrients are added; if too high, dilution or adjustments in nutrient concentrations are made.

☆ **Irrigation Scheduling:** Irrigation timing and frequency are adjusted based on both EC and pH measurements. For example, if EC is high, more frequent irrigation might be needed to leach excess salts.

Benefits

☆ **Precision:** Allows precise control over nutrient application, minimizing waste and optimizing plant uptake.

☆ **Optimization:** Ensures plants receive nutrients in the right proportions and at the right times, maximizing growth and yield.

☆ **Cost Efficiency:** Reduces the risk of over-fertilization and associated costs, as well as potential environmental impacts.

Practical Implementation

Equipment

☆ EC and pH sensors or meters.

☆ Automated control systems or manual adjustments based on sensor readings.

☆ Software for data analysis and decision support.

Calibration and Maintenance

☆ Regular calibration of sensors and meters to ensure accuracy.

☆ Maintenance of irrigation and fertigation equipment to prevent malfunctions that could affect nutrient delivery.

Integration with Other Practices

☆ Often integrated with other precision agriculture techniques such as soil moisture monitoring, weather forecasting, and crop modeling for comprehensive farm management.

Challenges

☆ **Complexity:** Requires understanding of plant nutrient needs and sensor technology.

☆ **Cost:** Initial investment in equipment and ongoing maintenance.

☆ **Skill Requirement:** Proper training and expertise for interpretation of data and making adjustments.

EC and pH-based irrigation/fertigation scheduling is a sophisticated approach that leverages real-time data to optimize nutrient management and water use efficiency in agriculture. It offers significant benefits in terms of crop productivity and resource conservation when implemented correctly.

Chapter 7

Canopy Management of Horticultural Crops

Introduction: An Overview

Canopy in a fruit tree refers to its physical composition comprising of stems, branches, shoot and leaves. The density of canopy is determined by the number and size of the leaves. Canopy architecture is determined by the number, length and orientation of the stem and shoots. Canopy management of the fruit trees deals with the development and maintenance of the structure in relation to size, shape, orientation of branches and light interception for the maximum productivity and quality. The basic concept in canopy management of a perennial tree to make the best use of land and the climate factors for an increased productivity in three dimensional approaches. Canopy management includes a range of techniques to after the position and the amount of leaves, shoot and fruits in space which determines, to a large extent the plant geometry structure including spatial distribution of leaf area and leaf orientation.

Canopy management refers an interrelation of the physiology underlying the relationship between vegetative growth and production. Thus, the ultimate goal of canopy management is to optimize carbon allocation in fruit sinks without disturbing growth and development in other part of the tree. The influence of temperature, light, humidity and tree vigour on the productivity and quality of fruit and manipulation of tree canopy through training system, pruning practices and use of growth retardants for the best utilization and harvest. In the last few years, significant development and strong formation have taken place in the development of tree canopy forms and new production system. High yield of good quality fruits produced under such systems are attributes to high light interception and distribution within the canopy.

Why We Need Canopy Management?

Light is critical to growth and development of trees and their fruits. The green leaves harvest the sunlight to produce carbohydrates and sugars which are transported to the sites where they are needed – buds, flowers and fruits. Better light penetration into the tree canopy improves tree growth, productivity, yield and fruit quality. The density and orientation of planting also impact light penetration in an orchard. Generally, in close planting, quicker shading becomes a problem.

Strong bearing branches tend to produce larger fruits. The problem of a fruit grower is initially to build up a strong and balanced framework of trees, then equip them with appropriate fruiting. Obviously, pruning in the early years has to be of a training type to provide strong and stocky framework with well-spaced limbs or any other desired shape.

Canopy Management

Canopy Management is the manipulation of tree canopies to optimise the production of quality fruits. The canopy management particularly affects the quality of sunlight which is intercepted by tree for proper tree shape and it determines the presentation of leaf area to enhance incoming radiation. Advancing knowledge in tree architecture, growth physiology, water management *etc.* has enabled farmers to adopt closer planting and maintaining reachable canopy by tailoring soil and crop management applications to fit in varying conditions in the field. It enables profitable cropping, high, regular yields and improved farm management practices, leading to higher productivity. This system of management produces high and regular yields of good quality fruits and low labour requirement to meet ever rising production costs.

Principle of Canopy Management

The aim is to induce trees to maintain a balance in between vegetative and fruiting wood, with uncrowded bearing wood situated conveniently within the relative shelter of the well-lit, adequately ventilated inner canopy.

The CMS (Canopy management strategy) manage tree complexity through three phases

- ☆ Induction of complexity in young trees.
- ☆ Maintenance of complexity in bearing trees.
- ☆ Reduction of complexity of large, old trees declining due to the effects of age and/or shading.

Basic principles in canopy management are:

- ☆ Maximum utilization of light.

☆ Avoidance of built-up microclimate which is congenial for diseases and pest infestation.

☆ Convenience in carrying out the cultural practices.

☆ Maximizing productivity with quality fruit production.

☆ Economy in obtaining the required canopy architecture.

Objectives of Canopy Management

Canopy management is an essential tree management operation starts from the first year of plant and enables the plant to produce/yield high quality and quantity by providing proper framework and more fruiting/yielding area.

☆ To remove the apical dominance for encouraging branching.

☆ To remove unproductive over crowded branches.

☆ To remove diseased and dead wood branches.

☆ To encourage vegetative growth.

☆ To control the overall size of the fruit tree.

☆ To regulate fruiting for regular cropping.

☆ To give particular training.

Requirement of Canopy Management

If canopies are trained and pruned there is reduction of water use. The more canopy reduction, the more transpiration reduction.

Importance of Canopy Management

☆ For proper density of trees/unit area

☆ For proper use of natural resources

☆ For better management of agronomical activities

☆ For good quality fruits

☆ Increase in production/unit area

☆ Annually regular fruit production

The main objectives of canopy management are given below:

☆ To get the higher yield with good quality.

☆ To maintain a good balance between root and shoot growth.

☆ Formation of strong crotches/crotch angle.

☆ To remove unwanted, overcrowding, dead disease and pest affected shoots.

☆ To regulate the tree architecture or form desire shape for high density planting system.

☆ To facilitate the management practices like spraying, harvesting *etc.*

☆ To proper utilize air, light and temperature efficiently.

☆ To regulate exposure of plant to light and air.

☆ To make accessibility to machinery between rows.

Principles of Canopy Management

☆ Maximum utilization of light.

☆ Avoidance of built-up microclimate congenial for diseases and pest infestation.

☆ Convenience in carrying out the cultural practices.

☆ Maximizing productivity with quality fruit production.

☆ Economy in obtaining the required canopy architecture.

☆ In many fruit crops, improved production and fruit quality has come from producing more fruit from smaller trees.

☆ Rejuvenation of declining in productivity and fruit quality in large over grown orchards.

☆ Small trees are better in capturing and converting sunlight in to fruit then large trees.

☆ Reduction in extra expense in harvesting at large trees.

☆ Safety risk for the harvest of bigger trees.

Mango

☆ Identify uprightly growing branches in each tree and thin out for increasing the productivity.

☆ Remove only one or two uprightly growing branches from center of tree to reduce tree height significantly and increase availability of light inside the canopy for better photosynthesis.

☆ Cutting of uprightly growing branches should be done during October-December from the base of their origin.

☆ During removal of branches, first cut should be given on lower side of branch to give a smooth cut and avoid bark splitting.

☆ Protect branches with wide crotch angle as they are more productive.

☆ In bearing mango trees, not more than 25 per cent biomass should be removed at a time for batter canopy management otherwise it results in excessive vegetative growth.

☆ Under high density planting system, remove 10-15 per cent biomass annually during October-December to increase light penetration inside the canopy. Removal of 10-15 per cent biomass should include crisscross branches, dead wood and diseased shoots.

Litchi

Management of optimum stature of litchi tree with compact and stereo bearing canopy is an important aspect of orchard management. Hence, giving proper shape to trees from initial stage and pruning of branches after harvesting in bearing trees is essential. Since more shoot sprouting takes place in young plants, proper shape is quite convenient at this stage.

- ☆ Single stem air layered plants should be raised in bags and allowed to grow up to 40-50 cm. The air layered plants have strong tendency to produce branches at the ground level which are pinched or pruned.

- ☆ Further, strong, well-spaced outshoots are allowed to form the main branches. It is necessary to continue shaping by removing all the branches forming crotches with main branches as and when they grow.

- ☆ To develop good and compact canopy, 25-30 cm fruit bearing shoots at the time of harvesting are removed. In this way, 2-3 new terminals develop which consequently develop into fruiting branches next season.

- ☆ Unproductive trees are pruned heavily to develop new fruitful shoots. In such cases, heavy reiterative pruning, usually up to limbs at a height of 4-5m is commonly followed, supplementary with heavy application of nutrients.

- ☆ These new shoots start fruiting 2-3 years after pruning. Thereafter, general pruning is followed to maintain ideal vigor and productivity of trees.

Mandarin (Citrus)

☆ The trees at planting time are headed back more severely to a height of 70-80 cm from the ground level.

☆ Pruning or cutting back of one year old shoots to half length (50 per cent of the total) or to full length is recommended for obtaining proper yield of high quality fruit. The pruning, therefore, is done to keep the balance between fruiting and vegetative growth.

☆ Pruning of some of the shoot certainly removes a part of fruiting area and helps maintain regular cropping. The dried up branches found in the lower part of the plant too are removed.

Lime (K. lime)

Acid lime plants may be trained to modified central leader system, with a smooth trunk up to 75-100 cm height from the ground level and 4-5 well spaced and well spread branches, as scaffolding branches.

- ☆ All sprouts appearing on the trunk up to a height of 75-100 cm should be removed. Similarly on grown up trees, the water suckers appearing on main trunk and scaffolding branches should be removed promptly.

- ☆ Once a young plant is trained to a desired shape, it requires very little pruning. Light pruning may be given during later years.

- ☆ Light pruned trees make more development of roots and shoots, producing fruit earlier that those pruned heavily Pruning of bearing trees though differ with variety, chiefly consists of removal of dead, dried, diseased, broken and cris-cross branches, whose existence is detrimental to the health of trees. Removal of water suckers is also essential.

- ☆ Pruning may be done just after harvesting. Soon after pruning, the cut ends may be smeared with Bordeaux paste or Blitox.

Guava

Untrained or unpruned guava trees become huge and unmanageable after a few years of growth. The bearing area is reduced and the interior of plants become entirely without fruits.

- ☆ Trees are topped to a uniform height of 60- 70 cm from the ground level, 2-3 months after planting to induce the emergence of new growth below the cut points.

☆ Three to four equally spaced shoots are retained around the stem to form the main scaffold limbs of tree. These shoots are allowed to grow for 4-5 months after topping until they attain a length of 40-50 cm.

☆ The selected shoots are further pruned to 50 per cent of their length for inducing multiple shoots from the buds below the cut end. Newly emerged shoots are allowed to grow up to 40-50 cm and pruned once again for emergence of new shoots. This is chiefly done to obtain the desired shape.

☆ The pruning operations continue during the second year after planting. After two years, short branches within the tree canopy produce a compact and strong structure. All the plants are confirmed to a hedge shape of 2 m inter row width and 2.5 m height for which pruning is performed in January and May-June every year.

Banana

☆ In most banana growing region, solar radiation is abundant and productivity of banana largely depends upon the efficient utilization of this resource. In multistory cropping system, banana is grown to harness maximum light, land and nutrient availability. Light interception, soil fertility, climatic conditions, soil moisture *etc.*

☆ Pruning of surplus leaves is a common operation in banana cultivation. It improves light penetration and reduces disease spreading through senescent leaves. For optimum crop production, minimum of 12 leaves are required to be retained.

Sapota

☆ Sapota is an evergreen tree so it requires no regular pruning but regulation of vegetative growth to improve productivity and quality of fruits. A seedling tree grows excellently giving a shape of an umbrella. Plants require training for appropriate shape and framework development.

☆ Most trees are trained in central leader system. During initial year, plants are topped to 60-70 cm above the ground level. After emergence of new shoots below the cut point, 3-4 well spaced scaffold limbs are selected and allowed to grow to make a strong framework.

Training

It means developing a desired shape of the tree with particular objectives by controlling habit of growth. Training is started from nursery stage of plant. Some fruit crops mostly like grape vines, ber, fig, guava *etc.* require training.

Training may be defined as the art and science of cutting away of portion of plant to improve its shape, to influence its growth, flowering and fruitfulness and to improve the quality of the product. It is done to divert a part of plant energy from one part to another part of plant or else it can be defined as judicial removal of plant parts to maintain the shape and size of plants to maintain the frame work of plant. Unmanaged tree canopy not only reduces the productivity of fruits but also detoriates the quality of produce as well. Good Canopy management practices results in good sunlight penetration and good yield. It enables profitable cropping, high, regular yields and improved farm management practices, leading to higher productivity. This system of management produces high and regular yields of good quality fruits and low labour requirement to meet ever rising production costs.

Objectives of Training

☆ Provides more light and air to the centre of the tree to expose maximum leaf surface to the sun.

☆ Proper growth of the tree so that various cultural operations such as spraying, harvesting can be performed easily and at lower cost.

☆ Protects the tree from sun burn and wind damage.

☆ Secures a balanced distribution of fruit bearing capacity of the tree.

Principles of Training

☆ Involves in formation of the main frame work, which must be strong. The branches must be suitable and spaced apart as well as the tree must be balanced on all the sides.

☆ It never allows several branches to grow at one place or very near each other.

☆ Another important point about training is that if two branches are growing at the same point try to train them to grow at a wider angle. Narrow angle is always weak.

Types of Training Systems

Central Leader System

In this system the central leader branches are allowed to grow independently so that it will grow more rapidly and vigorously than the side branches and tree become tall. Such tree bears fruit more near the upper portion. The lower branches are less vigorous and less fruitful.

Advantages

Such trees are structurally best suited to bear crop load and to resist the damage from strong winds.

Disadvantages

☆ Trees under this system grow too tall and are less spreading.

☆ The tree management (spraying, pruning, thinning and harvesting) is difficult.

☆ Impact of shading effect on interior canopy (the lower branches of such trees may be so much in shade that the fruit may not be able to develop proper colour.)

Open Centre or Vase System

The main stem is allowed to grow only up to a certain height about 1.5 to 1.8 m and then it cut for development of lateral branches. It allows full sunshine to reach each branch and resembles a vase like structure.

Advantages

☆ The trees trained allow maximum sunshine to reach the branches.

☆ Better clouration of fruits on the interior side of the tree.

☆ Trees are more fruitful and low spreading tree greatly facilitate operations like spraying, pruning, thinning and harvesting.

Disadvantages

Such trees are structurally weak, and their limbs are more likely to break with crop load and strong winds. This system does not only need severe pruning to start with but also constant effort to maintain its form through drastic pruning treatment.

Delayed Open Centre or Modified Leader System

It is intermediate between the above systems. It is developed by first training the tree to the leader type by allowing the central axil to grow unpruned for the first four or five years. Then central stem is headed back and lateral branches are allowed to grow as in the open centre system.

Advantages

- ☆ The branches are well distributed, allowing plenty of sunshine to reach the interior of the tree.
- ☆ The trees are structurally strong and not prone to limb breakage.
- ☆ Owing to limited height of trees, spraying, pruning and harvesting may be done easily.

Spindle Bush System

Modification of the dwarf pyramid or intermediate between a vertical cordon and a bush form. It differs from the dwarf pyramid, it has no specific arrangement of scaffold branches and forms the vertical cordon, here the fruit is borne on short branches, not directly on the main stem or trunk. Most important feature of this system is the laying down of lateral shoots in a horizontal position with little or no summer pruning. It is trained with or without support posts with a central leader straight and with many small fruiting branches. These branches are bent out and down by spreaders to develop wide crotches and to induce early fruiting. Tree spread is controlled by cutting back the shoots to ½ to ¾ of their length or back to weak laterals.

Overhead Trellis or Bower System

Bower system of training provides a desirable microclimate in the vine canopy and reduces the adverse effects of arid and hot weather on vine metabolism and life. In this system vines are spread over a pandal mounted at 2-2.4 m above the ground on poles made up of concrete, stone or iron.

Modified Bower or Telephone System

It is similar to bower system, here in every two meter as space is kept to walk and carry out cultural operations.

Cordon System

This is a system where in espalier is allowed with the help of training on wires. This system is followed in vines as these are incapable of standing on their stem.

This can be trained in single cordon or double cordon and commonly followed in crops like grape and passion fruit.

Kniffin System

A main stem or a vine is carried to the upper supporting wires. Renewed fruiting vines are tied. The bearing of fruiting vines are allowed to hang down. It is most adopted, easily understandable and easy to maintain. Provides higher yield and cost are effective in nature. The yield may vary hence it is important to adopt the specific system depending upon the crop.

Head Training Systems

When the growth commences, two shoots are selected and the remaining shoots are removed. The two shoots selected are then trained up using the stake and tied loosely to provide support but prevent girdling.

Advantages

☆ Ease of pruning to long canes.

☆ Vertical distribution of fruit.

☆ More compatible with tolerating winter injury than cordon systems.

Disadvantages

☆ Requires annual tying of canes.

☆ Difficult to maintain quality on lower wires (shading).

☆ Not compatible with systematic leaf removal and shoot positioning.

Pyramids Training Systems

Consists of a tree with a central stem about 2.5 m tall from which short branches radiate in successive tiers so that a pyramidal shape is build up. Fruiting spurs are developed on the short branches. Summer pruning forms an essential part of success with dwarf pyramids.

Espaliers

Similar to Kniffin system of training grapes. The trellis is 5 feet high with either 2 wires at 3 feet and 5 feet or 4 wires at 2, 3, 4, and 5 feet.

Y Shaped

One of the highest yielding systems. This system appears to be V shaped. Here trees are planted about 6 x 1 m apart (1668 trees/ha) bearing starts in the second year. Main framework consists

Advantages of Training Systems

Prevent weaker stem from lodging: Lodging means the displacement of the stem or root from their proper position. However, lodging is caused by many factors including wind and rain. And considering the weaker stem of plants and weight of the fruits, the right option is to provide support to get fruitful results.

Helps positioning of the leaves for adequate sunlight: Sunlight is very vital to the growth and increases the yield of crops. When crops do not receive an adequate amount of sunlight, they tend to wither or grow poorly. Trained crops are directed to receive an adequate amount of sunlight as they grow vertically. The leaves are better exposed to sunlight and the outcome is successful.

At a Glance

Unmanaged tree canopy not only reduces the productivity of fruits but also detoriates the quality of produce as well. Good Canopy management practices results in good sunlight penetration and good yield. It enables profitable cropping, high, regular yields and improved farm management practices, leading to higher productivity. This system of management produces high and regular yields of good quality fruits and low labour requirement to meet ever rising production costs.

Chapter 8

Remote Sensing and Geographical Information Systems

Introduction: An Overview

In recent times, the domain of Remote Sensing and GIS has gained considerable allure and appeal due to its quick growth and abundant prospects. Numerous organizations allocate substantial financial resources to these domains. The question arises as to why these fields have become increasingly relevant in recent years. There are two primary causes for this. In recent times, scientists, researchers, students, and even ordinary individuals are increasingly interested in gaining a deeper comprehension of our environment. The term "environment" refers to the geographical space of the study area and the occurrences that happen inside it. Put simply, we have recognized that geographic space, along with the accompanying data, is an integral part of our daily lives. Nearly every decision we make is influenced or determined by some aspect of geography. The progress in advanced space technology, which enables the acquisition of vast amounts of spatial data, combined with the decreasing costs of computer hardware and software capable of processing this data, has made Remote Sensing and Geographic Information Systems (G.I.S.) accessible not only for complex environmental and spatial scenarios but also for a growing number of people.

Milestones in History of Remote Sensing

Year	Milestones Achieved
1800	Discovery of Infrared by Sir W. Herchel
1839	Beginning of Practice of Photography
1847	Infrared Spectrum Shown by J.B.L. Foucault
1859	Photography from Balloons
1873	Theory of electromagnetic spectrum by J.C. Maxwell
1909	Photography from Airplanes
1916	World War I: Aerial Reconnaissance
1935	Development of Radar in Germany
1940	World War II: Application of Non-Visible parts of Electromagnetic radiation
1950	Military Research and Development
1959	First Space Photograph of the Earth (xplorer-6)
1960	First Meteorological Satellite launched
1970	Skylab Remote Sensing Observations from Space
1972	Launch LANDSAT-I and rapid advancement in digital image processing
1982	Launch od LANDSAT-4 with new generation of sensors (TM)
1986	French Commercial Earth Observational Satellite SPOT
1986	Development of hyperspectral sensors
1990	Developing high resolution space borne systems and first commercial developments in remote sensing
1998	Towards cheap one-goal satellite missions
1999	Launch EOS: NASA Earth observing mission
1999	Launch of IKONOS, very high spatial resolution sensor systems

Remote Sensing

It is a process of obtaining information and monitoring about an object, area, and the environment through the analysis of data acquired by a device without being in physical contact or direct touch. It involves the collecting and storing of spatial data of the environment without physical contact with the object by using EMW (Electromagnetic waves).

☆ The ability to collect information over large spatial areas.

☆ The record is unprejudiced one which is stored permanently.

☆ It allows for the collection of data over a variety of scales and resolutions,

☆ It can be used for crop identification, crop area, biomass and yield estimation.

Types of Remote Sensing

☆ **Active Remote Sensing:** Active Remote Sensing utilizes its own energy source for illumination. It releases energy to perform scans of items and places in order to gather data.

☆ **Passive Remote Sensing:** Passive Remote Sensing relies on external energy sources, such as the Sun, to detect naturally reflected or emitted electromagnetic radiation (EMR). Passive sensors capture and analyze the energy emitted by the sun when it is shining in order to gather information.

Remote Sensing and its Components

Remote sensing is the scientific process of gathering data about the Earth's surface without physical contact. This is accomplished through the detection and measurement of reflected or emitted energy, followed by the manipulation, examination, and utilization of that data. Remote sensing often involves the interaction between incident radiation and the targets of interest. Components of remote sensing is depicted in Figure 8.1.

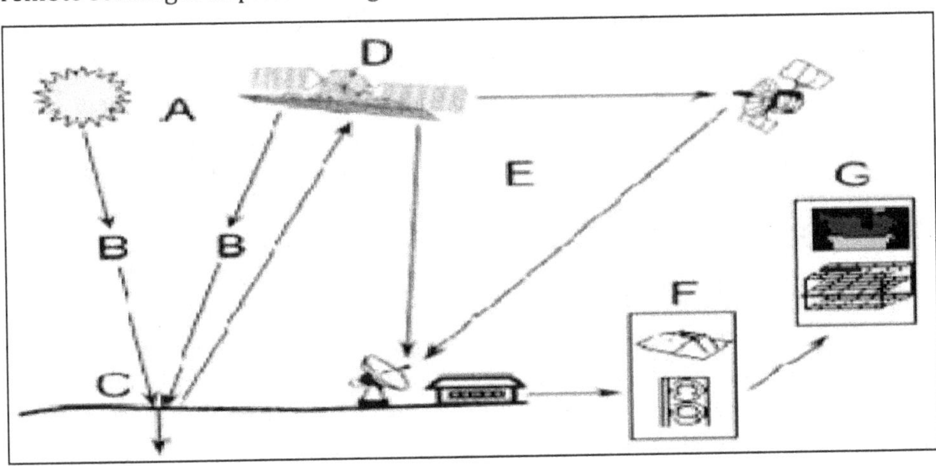

Figure 8.1. Components of Remote Sensing.

☆ **Energy Source or Illumination (A):** The primary prerequisite for remote sensing is the presence of an energy source that emits or supplies electromagnetic radiation to the desired target.

☆ **Radiation and the Atmosphere (B):** As the energy propagates from its origin to the destination, it will encounter and engage with the atmosphere it traverses. This contact may occur again as the energy propagates from the target to the sensor. Interaction with the Target (C) - once the energy makes its way to the target through the atmosphere, it interacts with the target depending on the properties of both the target and the radiation.

- ☆ **Recording of Energy by the Sensor (D):** Once the energy has been dispersed or released from the target, we need a sensor (which is not in direct touch with the target) to gather and document the electromagnetic radiation.

- ☆ **Transmission, Reception, and Processing (E):** The sensor's recorded energy must be communicated, typically in electronic format, to a receiving and processing unit where the data is transformed into an image, which can be either a hardcopy or a digital version.

- ☆ **Interpretation and Analysis (F):** The processed image is analyzed, either visually or digitally, to obtain information about the target that was lighted.

- ☆ **Application (G):** The culmination of the remote sensing process occurs when we utilize the extracted information from the imagery to enhance our comprehension of the target, unveil novel insights, or aid in resolving a specific issue.

Applications of Remote Sensing in Agriculture and Horticulture

Remote sensing is the method of gathering data about the Earth's surface by detecting the radiation it reflects and emits, without physically touching the thing. Remote sensing often involves the interaction between incident radiation and the targets of interest (Figure 8.2). The electromagnetic radiation that is most beneficial for distant sensing encompasses visible light (VIS), near infrared (NIR), shortwave infrared (SWIR), thermal infrared (TIR), and microwave bands (Figure 8.3). Passive remote sensing sensors detect and record the radiation that is reflected or released by objects. On the other hand, active sensors create their own radiation, which interacts with the target being studied and then returns to the measuring instrument.

Remote sensing has revolutionized the agricultural and horticultural industries by providing invaluable insights into crop health, environmental conditions, and land management practices. The applications of remote sensing in these sectors are vast and multifaceted, encompassing everything from yield prediction to disease detection. Remote sensing plays a crucial role in modern agriculture and horticulture by providing actionable insights that enhance productivity, sustainability, and resilience in agricultural systems. Its applications range from crop health monitoring and yield prediction to precision agriculture and environmental conservation, empowering farmers and researchers to make informed decisions for the future of food production.

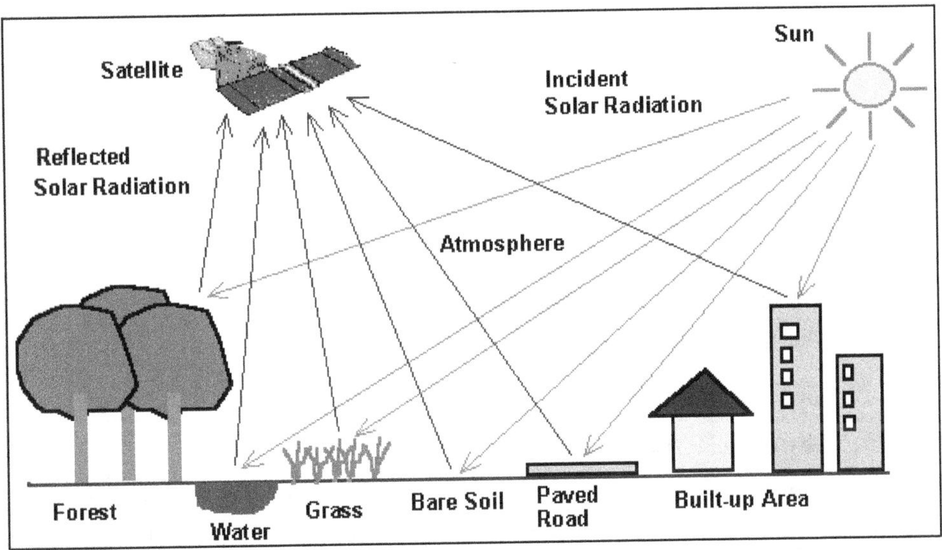

Figure 8.2. A Depiction of Processes Involves in Remote Sensing.

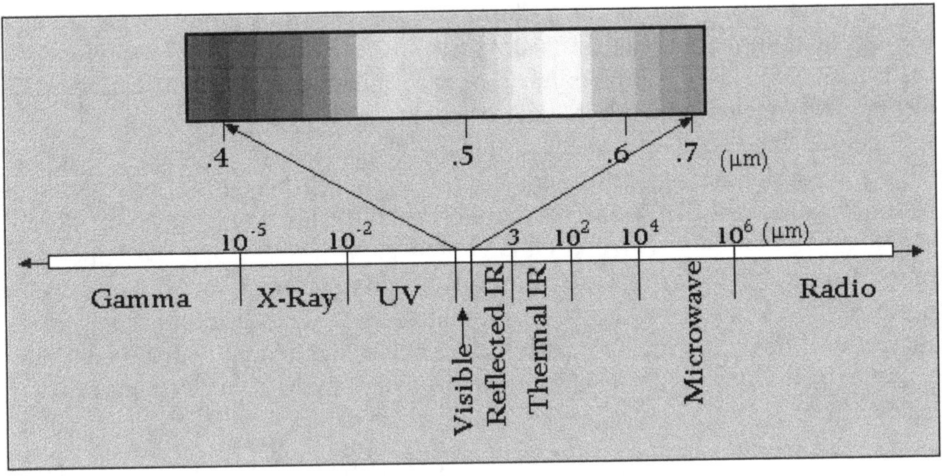

Figure 8.3. Full Range of Electromagnetic Spectrum.

IRS Data Applications

Indian Remote Sensing satellites are utilized for diverse purposes in the survey and management of resources. Pre-harvest crop area and production estimation of major crops.

- ☆ Drought monitoring and assessment based on vegetation condition.
- ☆ Flood risk zone mapping and flood damage assessment.
- ☆ Hydro-geo-morphological maps for locating underground water resources for drilling well.

☆ Irrigation command area status monitoring

☆ Snow-melt run-off estimates for planning water use in down stream projects

☆ Land use and land cover mapping

☆ Urban planning

☆ Forest survey

☆ Wetland mapping

☆ Environmental impact analysis

☆ Mineral Prospecting

☆ Coastal studies

Remote Sensing in Agriculture

According to a recent assessment by the FAO, the world population is projected to reach 9.15 billion by 2050. To meet the needs of this growing population, current food production will need to expand by 60 per cent. Various initiatives are currently being implemented to enhance overall production in order to meet the needs of the rapidly growing population. These efforts involve improving production efficiency through methods such as high intensity agriculture, optimized water usage, and the cultivation of high yield cultivars. Agricultural production has pronounced seasonal trends that are closely tied to the biological life cycle of crops. The production is influenced by the physical landscape, including factors such as soil type, as well as climate variables and agricultural management approaches. All of these variables exhibit significant spatial and temporal variability. Furthermore, since output can fluctuate rapidly as a result of unfavorable growth conditions, it is imperative for agricultural monitoring systems to operate in real time in order to maximize productivity. Hence, the utilization of remote sensing is essential for monitoring agricultural fields, assessing crop and soil health, managing water quality and quantity, and studying atmospheric conditions, all with a focus on maximizing crop output.

Over the past twenty years, remote sensing techniques have been used to investigate various agricultural applications, including distinguishing between different crops, estimating crop acreage, assessing crop conditions, estimating soil moisture, predicting crop yields, implementing precision agriculture, conducting soil surveys, managing agricultural water resources, and providing agro-meteorological and agro-advisory services. The utilization of remote sensing in agriculture, specifically in the study of crops and soils, is exceedingly intricate due to the constantly changing and inherently intricate nature of biological materials and soils (Myers, 1983). Remote-sensing technology offers numerous advantages compared to traditional approaches in agricultural resources survey. The benefits of this approach include: (a) the capacity to provide a comprehensive overview, (b) the possibility for rapid data collection, (c) the ability to repeatedly monitor and detect changes, (d) cost-effectiveness, (e) improved accuracy, and (f) the utilization of hyperspectral data to enhance the amount of information obtained. As previously said, remote sensing has numerous applications within the agriculture industry. Here is a synopsis of these applications.

Crop Production Forecasting

Remote sensing is used to forecast the expected crop production and yield over a given area and determine how much of the crop will be harvested under specific conditions. Researchers can be able to predict the quantity of crop in a given farmland over a given period. Crop production forecasting using remote sensing involves using satellite imagery and other remote sensing techniques to monitor crops' growth and health throughout the growing season. Here's an outline of how this process typically works:

- ☆ **Data Acquisition:** Satellite imagery is obtained from various sources such as NASA's MODIS, ESA's Sentinel satellites, or commercial satellite providers like Planet Labs. These images provide valuable information about the land surface, including vegetation indices, temperature, and moisture content.

- ☆ **Preprocessing:** Raw satellite data often requires preprocessing to correct for atmospheric distortions, cloud cover, and other artifacts. This step ensures that the data accurately reflects the conditions on the ground.

- ☆ **Image Analysis:** Once the data is preprocessed, various remote sensing techniques are applied to analyze the images. This may include calculating vegetation indices like NDVI (Normalized Difference Vegetation Index) or EVI (Enhanced Vegetation Index), which provide insights into crop health and vigor.

- ☆ **Crop Classification:** Remote sensing data can be used to classify different types of crops based on their spectral signatures. This step is essential for distinguishing between different crops in a region.

- ☆ **Yield Estimation:** By combining satellite imagery with crop models and historical data, it's possible to estimate crop yields for the current growing season. This involves extrapolating from the observed crop conditions to predict final yields at harvest time.

- ☆ **Validation:** The accuracy of the crop production forecasts is validated using ground-truth data, such as field surveys or yield measurements. This step helps to calibrate the models and ensure the reliability of the forecasts.

- ☆ **Forecasting and Reporting:** The final step involves generating crop production forecasts based on the remote sensing data and communicating this information to stakeholders such as farmers, policymakers, and commodity traders. These forecasts can help in decision-making related to crop management, food security, and market planning.

Remote sensing plays a crucial role in crop production forecasting by providing timely and accurate information about crop conditions over large geographic areas, helping to improve agricultural productivity and resilience in the face of climate variability and other challenges.

Assessment of Crop Damage and Crop Progress

In the event of crop damage or crop progress, remote sensing technology can be used to penetrate the farmland and determine exactly how much of a given crop has been damaged and the progress of the remaining crop in the farm. Remote sensing is invaluable in assessing crop damage and monitoring crop progress. Here's how it's done:

☆ **Initial Assessment:** When an event such as a storm, flood, or pest outbreak occurs, satellite imagery can provide a rapid overview of the affected area. This initial assessment helps identify the extent and severity of the damage.

☆ **Change Detection:** Remote sensing techniques allow for the comparison of current satellite images with historical data. By analyzing changes in vegetation cover, color, and texture, it's possible to identify areas where crops have been damaged or lost.

☆ **Vegetation Indices:** Vegetation indices like NDVI are particularly useful for monitoring crop health and progress. A sudden drop in NDVI values may indicate crop stress or damage, while a gradual increase suggests healthy growth.

☆ **Damage Classification:** Remote sensing data can be used to classify the type and severity of crop damage. For example, it can distinguish between waterlogged fields, areas affected by pests or diseases, and those damaged by extreme weather events.

☆ **Spatial Analysis:** GIS (Geographic Information Systems) techniques are often employed to analyze remote sensing data in combination with other spatial datasets, such as weather data and soil maps. This allows for a more comprehensive understanding of the factors contributing to crop damage and progress.

☆ **Temporal Analysis:** By monitoring changes in vegetation over time, remote sensing helps track crop development stages, such as planting, emergence, flowering, and maturity. This information is crucial for assessing crop progress and predicting harvest timing.

☆ **Decision Support:** Crop advisors, insurers, and policymakers use remote sensing data to make informed decisions about crop management, disaster response, and resource allocation. For example, satellite-based crop monitoring systems can trigger early warning alerts for farmers to take preventive measures against potential threats.

★ **Integration with Other Data Sources:** Remote sensing is often integrated with other data sources, such as ground surveys, drones, and weather stations, to improve the accuracy of crop damage assessments and progress monitoring.

In summary, remote sensing provides a powerful tool for assessing crop damage, monitoring crop progress, and supporting decision-making in agriculture. Its ability to provide timely and spatially explicit information helps mitigate risks, optimize resource allocation, and enhance food security.

Crop Identification

Remote sensing has played an important role in crop identification especially in cases where the crop under observation shows some mysterious characteristics. The crop data collected will be taken to labs where various aspects of crop including the crop culture are studied. Crop identification using remote sensing involves the use of satellite imagery and other remote sensing technologies to distinguish between different types of crops based on their spectral signatures and spatial characteristics. Here's an overview of the process:

★ **Data Acquisition:** Satellite imagery with high spatial and spectral resolution is acquired from sources such as NASA's MODIS, ESA's Sentinel satellites, or commercial providers like Planet Labs. The choice of imagery depends on factors such as the desired spatial resolution, temporal frequency, and spectral bands relevant for crop identification.

★ **Preprocessing:** Raw satellite data often undergoes preprocessing to correct for atmospheric distortions, cloud cover, and other artifacts. This step ensures that the data accurately represents the conditions on the ground and enhances the quality of subsequent analysis.

★ **Image Analysis:** Various remote sensing techniques are applied to analyze the satellite imagery and extract information about the land cover and land use. This may involve calculating spectral indices, such as NDVI (Normalized Difference Vegetation Index), which provide insights into the health and vigor of vegetation.

★ **Supervised Classification:** One common approach to crop identification is supervised classification, where the analyst provides training samples for each crop type of interest. Machine learning algorithms, such as Support Vector Machines (SVM) or Random Forests, are then trained to classify pixels in the satellite imagery into different crop classes based on their spectral characteristics.

★ **Unsupervised Classification:** Alternatively, unsupervised classification techniques can be used to identify clusters of pixels with similar spectral properties, which may correspond to different crop types. Clustering algorithms like K-means clustering or ISODATA can automatically group pixels into distinct classes without the need for training samples.

- ☆ **Object-Based Image Analysis (OBIA):** OBIA techniques take into account not only the spectral properties of individual pixels but also their spatial relationships and contextual information. This approach can improve the accuracy of crop identification by considering the spatial arrangement of crops within fields and incorporating additional ancillary data, such as terrain features or field boundaries.

- ☆ **Accuracy Assessment:** The accuracy of the crop identification results is typically evaluated using ground-truth data collected through field surveys or high-resolution aerial imagery. This validation step helps assess the reliability of the classification results and identify areas for improvement.

- ☆ **Integration with Ancillary Data:** Crop identification using remote sensing is often enhanced by integrating ancillary data sources, such as crop calendars, soil maps, and weather information. This additional contextual information can help refine the classification results and improve the accuracy of crop mapping.

Crop identification using remote sensing is a valuable tool for monitoring agricultural land use, supporting crop management decisions, and informing agricultural policy and planning efforts. By leveraging the capabilities of satellite imagery and advanced image analysis techniques, it enables the efficient and cost-effective mapping of crop distributions over large geographic areas.

Crop Acreage Estimation

Remote sensing has also played a very important role in the estimation of the farmland on which a crop has been planted. This is usually a cumbersome procedure if it is carried out manually because of the vast sizes of the lands being estimated. Crop acreage estimation using remote sensing involves using satellite imagery and other remote sensing data to measure the extent of agricultural land devoted to specific crops. Here's how the process typically works:

- ☆ **Data Acquisition:** High-resolution satellite imagery covering the agricultural area of interest is acquired from sources such as NASA's Landsat, ESA's Sentinel satellites, or commercial providers like DigitalGlobe. The imagery should capture the land during the growing season when crops are at their peak.

- ☆ **Preprocessing:** Raw satellite data undergoes preprocessing to correct for atmospheric effects, sensor artifacts, and geometric distortions. This step ensures that the imagery accurately represents the Earth's surface and enhances the quality of subsequent analysis.

- ☆ **Image Analysis:** Various remote sensing techniques are applied to analyze the satellite imagery and identify agricultural land and different crop types. This may involve image classification, segmentation, or object-based image analysis (OBIA) to delineate field boundaries and distinguish between different land cover classes.

☆ **Crop Classification:** Once agricultural land has been identified, supervised or unsupervised classification algorithms are used to classify pixels or image objects into different crop types. Training samples may be collected from ground-truth data or existing agricultural statistics to train the classification model.

☆ **Acreage Calculation:** After classifying the satellite imagery into different crop types, the total acreage of each crop can be calculated by summing the areas of pixels or image objects classified as that crop. This can be done at various spatial scales, from individual fields to larger administrative units such as counties or districts.

☆ **Accuracy Assessment:** The accuracy of the crop acreage estimation results is typically evaluated using ground-truth data collected through field surveys or high-resolution aerial imagery. This validation step helps assess the reliability of the classification results and identify areas for improvement.

☆ **Calibration and Validation:** Calibration may be necessary to account for factors such as differences in spectral signatures between crops, varying growing conditions, and changes in land use over time. Validation involves comparing the estimated acreage with ground-truth data or official agricultural statistics to ensure the accuracy of the results.

☆ **Reporting and Decision Support:** The final step involves generating reports and maps of crop acreage estimates and communicating this information to stakeholders such as farmers, policymakers, and agricultural researchers. These estimates can inform agricultural planning, resource allocation, and policy decisions.

Crop acreage estimation using remote sensing provides a valuable tool for monitoring agricultural land use, supporting crop management decisions, and informing agricultural policy and planning efforts. By leveraging the capabilities of satellite imagery and advanced image analysis techniques, it enables the efficient and cost-effective measurement of crop extents over large geographic areas.

Crop Yield Modelling and Estimation

Remote sensing also allows farmers and experts to predict the expected crop yield from a given farmland by estimating the quality of the crop and the extent of the farmland. This is then used to determine the overall expected yield of the crop. Crop yield modeling and estimation using remote sensing involves utilizing satellite imagery and other remote sensing data to predict and assess crop yields. Here's how the process typically unfolds:

☆ **Data Acquisition:** High-resolution satellite imagery, along with relevant ancillary data such as weather information, soil data, and historical yield records, are acquired for the study area. These data sources provide the necessary inputs for modeling crop growth and yield.

☆ **Preprocessing:** Raw satellite data undergoes preprocessing to correct for atmospheric effects, sensor artifacts, and geometric distortions. This step ensures that the imagery accurately represents the Earth's surface and enhances the quality of subsequent analysis.

☆ **Crop Growth Modeling:** Remote sensing data, combined with crop growth models such as the Crop Growth Monitoring System (CGMS) or the Decision Support System for Agrotechnology Transfer (DSSAT), are used to simulate the growth and development of crops throughout the growing season. These models take into account factors such as temperature, precipitation, solar radiation, soil characteristics, and crop management practices to predict crop growth and yield.

☆ **Calibration and Validation:** The crop growth models are calibrated and validated using ground-truth data collected from field surveys or experimental plots. This step ensures that the models accurately capture the relationships between remote sensing variables and crop yield under local conditions.

☆ **Yield Estimation:** Once the crop growth models are calibrated and validated, they can be used to estimate crop yields for the current growing season. Remote sensing data, such as vegetation indices (*e.g.*, NDVI), soil moisture, and temperature, are integrated into the models to provide spatially explicit information about crop conditions and potential yield variability across the study area.

☆ **Spatial Analysis:** GIS techniques are often employed to analyze remote sensing data in conjunction with other spatial datasets, such as land use/land cover maps, to assess the spatial distribution of predicted crop yields and identify areas of high and low productivity.

☆ **Uncertainty Analysis:** Uncertainty analysis is conducted to quantify the uncertainty associated with the yield estimates. This helps stakeholders understand the reliability of the predictions and identify sources of uncertainty that may need to be addressed.

☆ **Decision Support:** The final step involves communicating the crop yield estimates to stakeholders such as farmers, policymakers, and agricultural extension agents to support decision-making related to crop management, marketing, and food security. By providing timely and accurate information about crop yields, remote sensing-based yield modeling helps optimize agricultural production and resource allocation.

Crop yield modeling and estimation using remote sensing provide a powerful tool for monitoring crop productivity, assessing the impact of environmental factors on yields, and supporting evidence-based decision-making in agriculture. By integrating satellite imagery with crop growth models and ancillary data, it enables the efficient and cost-effective estimation of crop yields over large geographic areas.

Identification of Pests and Disease Infestation

Remote sensing technology aplays a significant role in identification of pests in farmland and gives data on the right pests control mechanism to get rid of the pests and diseases on the farm. Remote sensing can be a valuable tool for identifying pests and disease infestations in crops. Here's how it can be used for this purpose:

- ☆ **Spectral Signatures**: Different pests and diseases often cause characteristic changes in the spectral signatures of crops. Remote sensing data, such as multispectral or hyperspectral imagery, can be used to detect these spectral changes, allowing for the identification of pest and disease infestations.

- ☆ **Vegetation Indices**: Vegetation indices, such as the Normalized Difference Vegetation Index (NDVI) or the Enhanced Vegetation Index (EVI), can be calculated from remote sensing data to quantify the health and vigor of crops. Pests and diseases often lead to a reduction in vegetation indices due to damage or stress on the plants, making it possible to identify affected areas.

- ☆ **Temporal Analysis**: Remote sensing data collected over multiple time periods can be analyzed to monitor changes in crop health and detect anomalies indicative of pest and disease infestations. Time-series analysis allows for the identification of sudden changes or trends in vegetation indices that may be associated with pest outbreaks or disease spread.

- ☆ **Spatial Patterns**: Pests and diseases may exhibit spatial patterns within fields or across landscapes. Remote sensing data can be analyzed spatially to identify clusters or hotspots of vegetation stress that may indicate the presence of pests or diseases.

- ☆ **Integration with Ancillary Data**: Remote sensing data can be integrated with other data sources, such as weather data, soil maps, and crop phenology models, to improve the accuracy of pest and disease detection. For example, weather conditions may influence the spread of certain pests or diseases, so integrating weather data with remote sensing can provide valuable insights.

- ☆ **Machine Learning Algorithms**: Machine learning algorithms, such as supervised or unsupervised classification techniques, can be trained to automatically detect pest and disease infestations in remote sensing imagery. These algorithms learn to recognize patterns associated with pests and diseases and can be applied to large-scale datasets for automated detection.

- ☆ **Ground Truthing**: Ground truthing, which involves collecting field data to validate remote sensing results, is essential for confirming the presence of pests and diseases and calibrating detection algorithms. Field surveys, crop scouting, and sample collection can provide ground truth data that can be used to validate remote sensing-based assessments.

By leveraging the capabilities of remote sensing, it's possible to detect pest and disease infestations in crops more quickly and accurately than traditional methods alone. This early detection allows for timely interventions, such as targeted pesticide applications or disease management strategies, to minimize crop damage and maximize yields.

Soil Moisture Estimation

Soil moisture can be difficult to measure without the help of remote sensing technology. Remote sensing gives the soil moisture data and helps in determining the quantity of moisture in the soil and hence the type of crop that can be grown in the soil. Soil moisture estimation using remote sensing involves utilizing satellite or airborne sensors to measure and monitor the moisture content of the soil surface or the entire root zone. Here's how it's typically done:

☆ **Microwave Remote Sensing**: Microwave sensors are particularly useful for soil moisture estimation because microwave radiation can penetrate clouds and vegetation cover, allowing for measurements of soil moisture even under adverse weather conditions. Passive microwave sensors, such as those onboard NASA's Soil Moisture Active Passive (SMAP) satellite, measure natural microwave emissions from the Earth's surface, which are influenced by soil moisture content. Active microwave sensors, like those onboard ESA's Sentinel-1 satellites, emit microwave pulses and measure the backscattered signal, which is also sensitive to soil moisture.

☆ **Spectral Reflectance**: Optical remote sensing can also provide indirect measurements of soil moisture by analyzing the spectral reflectance of the Earth's surface. Near-infrared (NIR) and shortwave infrared (SWIR) bands are sensitive to changes in soil moisture, as wetter soils tend to reflect less light in these bands due to increased absorption by water. By analyzing multispectral or hyperspectral imagery, soil moisture can be estimated based on these spectral properties.

☆ **Thermal Infrared Remote Sensing**: Thermal infrared sensors measure the temperature of the Earth's surface, which is influenced by soil moisture content. Wetter soils tend to have lower surface temperatures due to the evaporative cooling effect of water. By comparing surface temperatures measured by thermal infrared sensors with meteorological data (*e.g.*, air temperature, humidity), soil moisture can be estimated.

☆ **Integration with Other Data**: Remote sensing data can be integrated with other sources of information, such as meteorological data, soil properties, and vegetation indices, to improve the accuracy of soil moisture estimation. For example, combining microwave remote sensing data with meteorological data allows for the development of empirical relationships (*e.g.*, soil moisture retrieval algorithms) that relate microwave brightness temperatures to soil moisture content.

☆ **Spatial and Temporal Resolution**: Remote sensing data can provide soil moisture estimates at various spatial and temporal resolutions, ranging from global-scale measurements with coarse spatial resolution to high-resolution measurements over smaller areas. The choice of sensor and data processing techniques depends on the specific application and requirements of the study.

☆ **Validation**: Validation of remote sensing-based soil moisture estimates is essential to assess their accuracy and reliability. Ground-based measurements collected from soil moisture sensors, in situ observations, or field surveys are commonly used for validation purposes. Statistical metrics such as correlation coefficients, root mean square error (RMSE), and bias are used to quantify the agreement between remote sensing estimates and ground truth measurements.

Remote sensing offers valuable tools for monitoring soil moisture over large spatial scales and providing essential information for various applications, including agriculture, hydrology, climate modeling, and natural resource management. By combining different remote sensing techniques and integrating them with ancillary data, it's possible to derive accurate and timely estimates of soil moisture content, which are crucial for understanding and managing Earth's water cycle.

Soil Mapping

Soil mapping is one of the most common yet most important uses of remote sensing. Through soil mapping, farmers are able to tell which soils are ideal for which crops and which soil require irrigation and which ones do not. This information helps in precision agriculture. Soil mapping using remote sensing involves the use of satellite or airborne sensors to collect data on soil properties and characteristics over large geographic areas. Here's how the process typically works:

☆ **Spectral Signatures**: Different soil types have distinct spectral signatures that can be detected and analyzed using remote sensing data. Optical sensors measure the reflectance of sunlight across different wavelengths, providing information about the composition and properties of the soil surface.

☆ **Multispectral and Hyperspectral Imaging**: Multispectral sensors capture imagery in several discrete bands across the electromagnetic spectrum, while hyperspectral sensors collect data with hundreds of narrow spectral bands. By analyzing the spectral signatures of soils in these images, it's possible to identify and map different soil types based on their unique reflectance patterns.

- ✰ **Vegetation Indices**: Vegetation indices derived from remote sensing data, such as the Normalized Difference Vegetation Index (NDVI) or the Soil Adjusted Vegetation Index (SAVI), can be used to infer soil properties indirectly. For example, NDVI values may vary depending on the soil's fertility, moisture content, and texture, allowing for the mapping of soil properties based on vegetation patterns.

- ✰ **Texture and Composition**: Remote sensing data can also be used to map soil texture and composition, such as clay, silt, and sand content. Different soil types exhibit characteristic spectral responses that can be correlated with their texture and composition, allowing for the identification and mapping of soil properties at a regional scale.

- ✰ **Terrain Analysis**: Terrain features derived from remote sensing data, such as slope, aspect, and elevation, can provide additional information about soil properties and distribution. For example, soil types may vary with topographic characteristics, with certain soils being more prevalent in low-lying areas or on steep slopes.

- ✰ **Machine Learning Algorithms**: Machine learning algorithms, such as Random Forests or Support Vector Machines, can be trained to classify remote sensing data into different soil types based on their spectral characteristics. These algorithms learn to recognize patterns in the data and can be applied to large-scale soil mapping efforts.

- ✰ **Integration with Ground Truth Data**: Ground truth data, such as soil samples collected from field surveys or soil maps generated from ground-based measurements, are essential for calibrating and validating remote sensing-based soil maps. These data are used to train classification algorithms and assess the accuracy of the resulting soil maps.

- ✰ **Spatial Resolution**: The spatial resolution of remote sensing data influences the level of detail and accuracy of soil maps. Higher-resolution imagery can capture finer-scale soil variations but may be limited in coverage, while lower-resolution imagery provides broader coverage but may lack detail.

Overall, remote sensing offers valuable tools for mapping soil properties and characteristics over large spatial scales, providing essential information for various applications, including agriculture, land use planning, environmental management, and natural resource conservation. By combining different remote sensing techniques with ground truth data and advanced analytical methods, it's possible to derive accurate and detailed soil maps that support informed decision-making and sustainable land management practices.

Monitoring of Droughts

Remote sensing technology is used to monitor the weather pattern of a given area. The technology also monitors drought patterns of the area too. The information can be used to predict the rainfall patterns of an area and also tell the time difference between the current rainfall and the next rainfall which helps to keep track of the drought. Remote sensing technology plays a crucial role in monitoring droughts by providing timely and spatially explicit information about various drought indicators. Here's how remote sensing is used for drought monitoring:

- ☆ **Vegetation Health Monitoring**: Remote sensing data, such as satellite imagery, can be used to monitor changes in vegetation health, which is often one of the earliest and most visible indicators of drought. Vegetation indices like the Normalized Difference Vegetation Index (NDVI) or the Enhanced Vegetation Index (EVI) are calculated from satellite imagery to assess the greenness and vigor of vegetation over time. A decrease in vegetation indices indicates stress due to water scarcity, which can be indicative of drought conditions.

- ☆ **Soil Moisture Estimation**: Remote sensing data, particularly from microwave and thermal infrared sensors, can provide information about soil moisture content, which is critical for assessing drought severity. Changes in soil moisture levels can be detected and monitored using satellite-based observations, helping to identify areas experiencing moisture deficits associated with drought.

- ☆ **Temperature Monitoring**: Remote sensing data can be used to monitor land surface temperatures, which can increase during periods of drought due to reduced soil moisture and vegetation cover. Thermal infrared sensors measure the thermal radiation emitted by the Earth's surface, allowing for the detection of anomalous temperature patterns associated with drought conditions.

- ☆ **Water Bodies Monitoring**: Remote sensing technology enables the monitoring of water bodies, such as lakes, reservoirs, and rivers, to assess changes in water levels and availability. Satellite imagery can be used to track changes in water extent, surface area, and volume over time, providing valuable information about the impact of drought on water resources.

- ☆ **Evapotranspiration Estimation**: Remote sensing data can be used to estimate evapotranspiration rates, which represent the combined loss of water from the soil and vegetation through evaporation and transpiration. By monitoring changes in evapotranspiration rates, it's possible to assess the water balance of an area and identify regions experiencing drought stress.

- ☆ **Drought Severity Index**: Remote sensing data can be integrated with meteorological data and ground-based observations to develop drought severity indices, such as the Standardized Precipitation Index (SPI) or the Vegetation Condition Index (VCI). These indices provide quantitative measures of drought severity and can be derived from satellite imagery and other remote sensing data sources.

☆ **Early Warning Systems**: Remote sensing technology enables the development of early warning systems for drought monitoring and forecasting. By integrating remote sensing data with climate models and hydrological simulations, it's possible to predict drought onset, duration, and severity, allowing for proactive mitigation measures and resource allocation.

Overall, remote sensing technology provides valuable tools for monitoring droughts by providing timely, accurate, and spatially explicit information about various drought indicators. By leveraging satellite imagery, sensor networks, and advanced analytical techniques, remote sensing helps governments, policymakers, and resource managers assess drought conditions, monitor trends, and implement effective drought response strategies.

Water Resources Mapping

Remote sensing is instrumental in the mapping of water resources that can be used for agriculture over a given farmland. Through remote sensing, farmers can tell where water resources are available for use over a given land and whether the resources are adequate. Water resources mapping using remote sensing involves the use of satellite imagery and other remote sensing data to assess and monitor various aspects of water resources, including surface water bodies, groundwater, and water quality. Here's how remote sensing technology is utilized for water resources mapping:

☆ **Surface Water Mapping**: Remote sensing data can be used to map and monitor surface water bodies such as lakes, rivers, reservoirs, and wetlands. Optical sensors on satellites capture imagery in the visible and near-infrared spectrum, allowing for the identification and delineation of water bodies based on their spectral signatures. Additionally, radar sensors on satellites, such as those on ESA's Sentinel-1 mission, can penetrate clouds and vegetation cover to detect water surfaces, making them useful for mapping surface water even under adverse weather conditions.

☆ **Water Level Monitoring**: Remote sensing technology enables the monitoring of water levels in lakes, rivers, and reservoirs over time. By analyzing satellite imagery acquired at different time periods, it's possible to track changes in water extent and volume, helping to assess water availability and manage water resources effectively. Radar altimetry and interferometric synthetic aperture radar (InSAR) techniques can also be used to measure water levels with high precision from space.

☆ **Groundwater Exploration**: Remote sensing data can provide valuable information for groundwater exploration and assessment. Thermal infrared sensors measure the temperature of the Earth's surface, which can be influenced by the presence of groundwater. Anomalies in surface temperature patterns may indicate the presence of subsurface groundwater flow or discharge areas. Additionally, radar and optical sensors can be used to identify geological features associated with potential groundwater resources, such as faults, fractures, and geological formations.

- ☆ **Water Quality Monitoring**: Remote sensing technology can be used to assess water quality parameters such as turbidity, chlorophyll concentration, and dissolved organic matter. Optical sensors on satellites measure the spectral reflectance of water bodies, allowing for the detection of changes in water color and clarity associated with variations in water quality. Multispectral and hyperspectral imagery can be analyzed to derive water quality indicators and assess the health of aquatic ecosystems.

- ☆ **Wetland Mapping and Monitoring**: Wetlands play a critical role in regulating water flow, filtering pollutants, and providing habitat for wildlife. Remote sensing data can be used to map and monitor wetland extent, vegetation cover, and hydrological dynamics. Optical and radar sensors provide information about wetland morphology, vegetation density, and inundation patterns, allowing for the assessment of wetland health and function over time.

- ☆ **Hydrological Modeling**: Remote sensing data can be integrated with hydrological models to simulate water flow, runoff, and recharge processes within watersheds. By combining satellite imagery with hydrological data and ground-based observations, it's possible to develop accurate models of water resources dynamics and assess the impacts of land use changes, climate variability, and human activities on water availability and quality.

Remote sensing technology provides valuable tools for mapping and monitoring water resources at regional and global scales, supporting informed decision-making and sustainable management of water resources. By leveraging satellite imagery, sensor networks, and advanced analytical techniques, remote sensing helps governments, policymakers, and resource managers assess water availability, monitor changes over time, and implement effective water resource management strategies.

Remote Sensing in Horticulture

Remote sensing and other geo-informatics technologies are an advanced technique used for precision and hi-tech farming these days. It helps to know the accurate condition of a crop and with what all stresses our crops are going through. Remote sensing plays an important role in the forecasting and predicting of drought, flood or any other natural calamities and also includes biotic stress like disease and pest. And how and when fertilizers and pesticides should be applied and at what ratio and amount should be applied to control and manage the situation. Remote sensing has solved most or our problems like collecting data, storing and analyzing *etc.* It also makes us easy to maintain a record and form a statistical report of all-season crops.

Horticulture is a crucial aspect of our daily lives, encompassing many crops such as fruits, vegetables, medicinal plants, fragrant plants, spices, and plantation crops. It influences our morning tea, clothing, and all our meals. As a result of the growing population and rising demand, there is a need for an increased food supply to adequately feed our population. However, the lack of moisture, nutrition, or sufficient labor prevents us from obtaining an abundance of these things, resulting in higher retail prices that are unaffordable for lower-middle-class families. In order to satisfy the needs of the people, hi-tech farming is a prominent and effective alternative. Remote Sensing can be combined with other sophisticated methods like Global Positioning System (GPS) and Geographical Information System (GIS). These strategies are significantly influencing the evaluation and control of horticultural activities. The application of agricultural meteorology is necessary for the successful planning and implementation of new technologies in crop cultivation. An agricultural weather and climate data system is crucial for expediting the production of goods, analysis, and predictions that impact decisions related to horticultural crops and management, irrigation scheduling, commodities trading, and marketing (Saxena, *et al.,* 2017).

Applications of Remote Sensing in Horticulture

Remote sensing technology has revolutionized horticulture by offering innovative tools for monitoring, managing, and enhancing various aspects of crop production. From precision agriculture to environmental monitoring, remote sensing plays a crucial role in optimizing crop yields and resource utilization in horticultural practices. By utilizing satellites, drones, and other advanced imaging technologies, remote sensing enables farmers and researchers to assess crop health, detect diseases and pests, monitor soil moisture levels, and evaluate environmental conditions. This data-driven approach allows for timely interventions, such as targeted irrigation, precision fertilization, and early pest detection, ultimately leading to improved crop quality, increased yields, and sustainable farming practices in the horticulture industry.

The first event of using remote sensing technique in India was documented during coconut wilting in 1970.

Crop Classification

The multispectral image of Remote sensing system plays an important role to determine the difference between different horticulture crops (like grasses, herbs, shrubs, trees and climbers) or whether it is a flourishing plant or an infected unhealthy plant/weed (Dakshinamurti *et al.,* 1971). Crop classification, a fundamental task in agricultural monitoring and management, involves identifying and categorizing different crop types within a given area. Remote sensing, with its ability to capture large-scale, high-resolution imagery of the Earth's surface, plays a pivotal role in automating and improving the accuracy of crop classification processes.

Remote sensing facilitates crop classification through various methods:

- ☆ **Spectral Signatures:** Different crops exhibit unique spectral reflectance patterns across the electromagnetic spectrum. Remote sensing sensors, such as multispectral and hyperspectral imagers, capture this information, allowing for the differentiation of crops based on their distinct spectral signatures. By analyzing the reflectance values at specific wavelengths, algorithms can classify crops effectively.

- ☆ **Temporal Analysis:** Remote sensing platforms provide frequent and repetitive coverage of agricultural areas, allowing for temporal analysis of crop growth stages and phenological changes. This temporal information, combined with spectral data, enhances the accuracy of crop classification by considering the dynamic nature of crop development over time.

- ☆ **Spatial Patterns:** Remote sensing imagery enables the extraction of spatial patterns and structural characteristics of crops, such as canopy density, shape, and arrangement. These spatial features serve as valuable input variables for crop classification algorithms, facilitating the discrimination between different crop types based on their spatial characteristics.

- ☆ **Machine Learning Algorithms:** Advanced machine learning techniques, such as support vector machines (SVM), random forests, and convolutional neural networks (CNN), are commonly employed for crop classification using remote sensing data. These algorithms learn complex patterns and relationships from labeled training data, enabling accurate and automated classification of crops across large agricultural landscapes.

The role of remote sensing in crop classification extends beyond accurate mapping of crop types. It facilitates agricultural decision-making by providing timely and spatially explicit information on crop distribution, acreage estimation, and land use changes. This information is invaluable for crop monitoring, yield forecasting, resource allocation, and policy formulation, ultimately contributing to sustainable agricultural practices, food security, and economic development.

Crop Insurance

Due to excess pollution and global warming, we never know how and when the climate will show its destructive effect. But now we can know prior, about weather conditions and take safety measures accordingly with the help of remote sensing. Crop insurance is a crucial risk management tool for farmers, providing financial protection against crop losses due to adverse weather conditions, pests, diseases, or other unavoidable circumstances. Remote sensing technology plays a significant role in improving the efficiency and effectiveness of crop insurance programs by enhancing the accuracy of risk assessment, claims processing, and overall decision-making processes. Here's how remote sensing contributes to crop insurance:

☆ **Risk Assessment:** Remote sensing data, including satellite imagery and aerial surveys, provide valuable information on crop health, growth stages, and spatial distribution. By analyzing this data, insurance companies can assess the vulnerability of different crops to various risks, such as drought, floods, or pest infestations. Remote sensing helps insurers identify high-risk areas and tailor insurance policies accordingly, ensuring adequate coverage for farmers facing the greatest threats.

☆ **Yield Estimation:** Remote sensing enables the estimation of crop yields by monitoring vegetation indices, such as NDVI (Normalized Difference Vegetation Index), throughout the growing season. These indices serve as indicators of crop health and productivity, allowing insurers to predict potential yield losses due to adverse events. Accurate yield estimation helps insurance companies set appropriate coverage levels and premiums, ensuring fair compensation for farmers in the event of crop failure.

☆ **Damage Assessment:** In the event of a crop loss, remote sensing facilitates rapid and objective damage assessment. High-resolution satellite imagery and aerial surveys enable insurers to quickly evaluate the extent and severity of crop damage caused by natural disasters or other hazards. This timely assessment ensures prompt claims processing and reduces the administrative burden associated with traditional, on-site inspections.

☆ **Fraud Detection:** Remote sensing technology can also help detect fraudulent insurance claims by comparing pre- and post-event imagery to verify the accuracy of reported losses. Inconsistencies between observed damage and claimed losses can raise red flags, prompting further investigation and deterrence of fraudulent activities.

☆ **Monitoring and Compliance:** Remote sensing enables continuous monitoring of insured agricultural areas throughout the policy period, allowing insurers to verify compliance with policy conditions, such as planting dates, crop types, and land use practices. This proactive monitoring helps mitigate moral hazard and adverse selection, ensuring the integrity of crop insurance programs and equitable treatment of policyholders.

The integration of remote sensing technology into crop insurance enhances risk management capabilities, improves decision-making processes, and ultimately strengthens the resilience of agricultural communities against unforeseen challenges and uncertainties. By leveraging remote sensing data, insurers can optimize insurance products, enhance customer satisfaction, and promote sustainable agricultural practices.

Crop Area Estimation

It helps to get the estimation of a particular crop cultivated in an area to support crop forecasting system at a regional level. Horticultural crops usually face fluctuation both in its production and consumption (Nageswara Rao, *et al.,* 2004). That's why genuine statistics concerning the area and production of horticulture crops is necessary for market planning and its export. Crop area estimation is a critical task in agricultural management, market analysis, and policy planning, providing valuable information on the spatial distribution and extent of different crops cultivated within a region. Remote sensing technology plays a pivotal role in enabling accurate and efficient crop area estimation through its capability to capture large-scale, high-resolution imagery of agricultural landscapes. Here's how remote sensing contributes to crop area estimation:

- ☆ **Satellite Imagery:** Remote sensing satellites equipped with multispectral and optical sensors capture detailed imagery of the Earth's surface at various spatial and temporal resolutions. These satellite images provide comprehensive coverage of agricultural areas, allowing for the identification and delineation of different crop types based on their spectral characteristics, such as colour, texture, and spatial patterns.

- ☆ **Image Processing Techniques:** Remote sensing data undergoes advanced image processing techniques to extract relevant information for crop area estimation. These techniques include image classification, segmentation, and feature extraction, which enable the differentiation and mapping of distinct crop types within the imagery. Supervised and unsupervised classification algorithms, coupled with machine learning approaches, help automate and improve the accuracy of crop classification processes.

- ☆ **Temporal Analysis:** Temporal analysis of satellite imagery over multiple growing seasons enables the monitoring of crop dynamics and phenological changes throughout the agricultural calendar. By analyzing temporal trends in vegetation indices, such as NDVI (Normalized Difference Vegetation Index), remote sensing facilitates the identification of planting dates, crop emergence, growth stages, and harvest timing. This temporal information is essential for accurately estimating crop area and assessing inter-annual variability in crop production.

- ☆ **Integration with Geographic Information Systems (GIS):** Remote sensing data is often integrated with GIS platforms to spatially organize, analyze, and visualize crop distribution patterns. GIS-based tools and spatial modeling techniques facilitate the aggregation of pixel-level crop classifications into larger geographic units, such as administrative boundaries or agricultural land parcels. This integration enhances the scalability and applicability of crop area estimation methods, allowing for regional or national-level assessments of agricultural land use.

☆ **Validation and Calibration:** Remote sensing-based crop area estimation models are validated and calibrated using ground-truth data collected through field surveys, agricultural statistics, and farmer interviews. This validation process ensures the accuracy and reliability of remote sensing-derived crop area estimates, while also identifying potential sources of error or uncertainty. Ground-truth validation data are essential for calibrating classification algorithms, correcting for misclassifications, and improving the overall performance of crop area estimation models.

In summary, remote sensing technology revolutionizes crop area estimation by providing timely, spatially explicit, and objective information on agricultural land use. By leveraging satellite imagery, advanced image processing techniques, and temporal analysis, remote sensing enables efficient and accurate mapping of crop extent, supporting informed decision-making in agricultural planning, resource allocation, and policy formulation.

Yield Monitoring System

Remote sensing is an important tool to estimate the yield of all seasonal and nonseasonal crops annually. It also helps us to know that production is enough to meet the needs and demands of the population. Yield monitoring systems are essential tools in modern agriculture for assessing crop productivity, optimizing inputs, and maximizing yields. These systems utilize various technologies to measure and record crop yields during harvesting operations. Remote sensing plays a crucial role in enhancing the accuracy and efficiency of yield monitoring systems by providing valuable spatial and temporal information on crop growth and variability across agricultural fields. Here's how remote sensing contributes to yield monitoring:

☆ **Vegetation Indices:** Remote sensing sensors, such as multispectral and hyperspectral imagers mounted on satellites or drones, capture spectral information from agricultural fields. Vegetation indices, such as NDVI (Normalized Difference Vegetation Index), NDRE (Normalized Difference Red Edge), and EVI (Enhanced Vegetation Index), derived from remote sensing data, serve as proxies for crop biomass and health. By analyzing changes in vegetation indices over time, farmers can assess crop growth and vigor, which are key indicators of potential yield.

☆ **Spatial Variability Analysis:** Remote sensing provides detailed spatial information on crop variability within fields. High-resolution satellite imagery or drone-based surveys enable the identification of areas with different levels of vegetation vigor, moisture content, and nutrient availability. This spatial variability analysis helps farmers understand the factors contributing to yield variations across their fields, allowing for targeted management interventions to improve overall productivity.

☆ **Yield Prediction Models:** Remote sensing data, combined with ground-based measurements and weather data, can be integrated into yield prediction models. These models utilize machine learning algorithms or statistical techniques to correlate remote sensing-derived vegetation indices with actual yield data collected from harvesters. By calibrating and validating these models over multiple growing seasons, farmers can forecast crop yields with greater accuracy, enabling proactive decision-making in crop marketing, storage, and logistics.

☆ **Harvest Monitoring:** During harvesting operations, remote sensing technology can be used to monitor the progress and efficiency of harvesters in real-time. Sensors mounted on harvesting equipment, such as yield monitors and GPS receivers, record georeferenced yield data as crops are harvested. These data are then integrated with remote sensing imagery to map spatial variations in yield across fields, identify potential yield-limiting factors, and optimize harvesting strategies for maximum efficiency.

☆ **Post-Harvest Analysis:** Remote sensing continues to play a role in post-harvest analysis by providing imagery for assessing crop residue cover, soil compaction, and field conditions after harvest. This information helps farmers evaluate the effectiveness of their harvest operations, identify areas requiring remediation or conservation measures, and plan for subsequent cropping cycles.

Remote sensing technology enhances yield monitoring systems by providing timely, accurate, and spatially explicit information on crop growth, variability, and productivity. By integrating remote sensing data into yield monitoring workflows, farmers can make informed decisions to optimize input use, mitigate risks, and maximize yields, contributing to sustainable and profitable agriculture.

Detecting Pest and Disease

Pest and Disease both responsible for the damage of crops and hence, affects the economic importance of horticulture crops. It has been demonstrated that remote sensing helps to identify the pest and disease and its damage stage (Usha *et al.,* 2013). Detecting pests and diseases early is crucial for minimizing crop damage and maximizing yields in agriculture. Remote sensing technology plays a significant role in pest and disease detection by providing timely and spatially explicit information on crop health and stress indicators. Here's how remote sensing contributes to detecting pests and diseases:

☆ **Spectral Signatures:** Remote sensing sensors capture electromagnetic radiation reflected or emitted by crops, which varies based on their physiological status and health condition. Pests and diseases cause changes in crop reflectance and emission spectra, altering their spectral signatures compared to healthy vegetation. By analyzing spectral data, such as NDVI (Normalized Difference Vegetation Index) or specific spectral bands sensitive to stress indicators, remote sensing can detect subtle changes associated with pest infestations or disease outbreaks.

☆ **Symptom Mapping:** Remote sensing imagery provides detailed spatial information on crop conditions across agricultural fields. By visually interpreting satellite or drone imagery, agricultural experts can identify visual symptoms of pest damage or disease infections, such as discoloration, wilting, lesions, or defoliation. These symptom maps help pinpoint areas of crop stress for further investigation and management interventions.

☆ **Temporal Analysis:** Remote sensing facilitates temporal analysis of crop health over multiple growing seasons, allowing for the monitoring of pest and disease dynamics over time. By comparing historical and current imagery, farmers and researchers can track the progression of pest infestations or disease outbreaks, assess their severity, and anticipate potential impacts on crop yields. Temporal analysis also helps differentiate between transient stress events and persistent pest or disease pressure.

☆ **Spatial Patterns:** Remote sensing enables the extraction of spatial patterns and distribution of pests and diseases within agricultural landscapes. By analyzing spatial variability in crop health indicators derived from satellite or drone imagery, such as NDVI gradients or texture metrics, researchers can identify hotspots of pest activity or disease prevalence. These spatial patterns inform targeted scouting efforts and pest management strategies, optimizing resource allocation and treatment efficacy.

☆ **Integration with Geospatial Technologies:** Remote sensing data can be integrated with geographic information systems (GIS) and other geospatial technologies to enhance pest and disease detection workflows. GIS-based tools facilitate the spatial analysis, visualization, and interpretation of remote sensing-derived information in the context of environmental factors, crop management practices, and landscape features. This integration enables comprehensive pest and disease risk assessment, decision support, and early warning systems for farmers and agricultural stakeholders.

Remote sensing technology provides a valuable tool for detecting pests and diseases in agriculture by leveraging spectral, spatial, and temporal information to identify crop stress indicators associated with pest infestations or disease infections. By integrating remote sensing data into pest and disease management strategies, farmers can implement timely and targeted interventions to mitigate crop losses, minimize pesticide use, and ensure sustainable crop production.

Monitoring Abiotic Stress

Remote sensing also helps to detect abiotic stresses like drought, flooding, salinity, temperature fluctuations *etc.* with its feature of hyperspectral and multispectral imaging. Remote sensing plays a crucial role in monitoring abiotic stress factors affecting crops, such as drought, salinity, nutrient deficiencies, and temperature extremes. Abiotic stresses can significantly impact crop growth, productivity, and resilience, making their timely detection and management essential for sustainable agriculture. Here's how remote sensing contributes to monitoring abiotic stress:

- ☆ **Spectral Signatures:** Remote sensing sensors capture electromagnetic radiation reflected or emitted by crops, which varies based on their physiological status and environmental conditions. Abiotic stresses alter crop spectral signatures by affecting chlorophyll content, leaf water content, and other biochemical properties. By analyzing spectral data, such as vegetation indices (*e.g.*, NDVI, NDWI, and EVI), remote sensing can detect and quantify the severity of abiotic stressors, such as water scarcity, soil salinity, or nutrient deficiencies.

- ☆ **Thermal Imaging:** Thermal infrared sensors mounted on satellites or drones measure the temperature of crops and soil surfaces, providing valuable information on heat stress and water status. Abiotic stresses, such as drought or heatwaves, increase crop temperature and alter thermal gradients within agricultural fields. Thermal imaging allows for the detection of temperature anomalies indicative of stress conditions, enabling early intervention and irrigation scheduling to mitigate heat-related crop damage.

- ☆ **Water Stress Monitoring:** Remote sensing facilitates the assessment of water stress in crops by measuring vegetation moisture content and water availability in soil. Sensors sensitive to water absorption bands, such as near-infrared and shortwave infrared wavelengths, capture changes in crop reflectance related to water stress. By analyzing these spectral responses, remote sensing can estimate crop water content, detect drought stress, and identify areas requiring irrigation or soil moisture management interventions.

- ☆ **Salinity Mapping:** Remote sensing imagery enables the mapping and monitoring of soil salinity levels across agricultural landscapes. Saline soils exhibit distinct spectral characteristics, such as high reflectance in the visible and near-infrared regions due to salt deposits. Remote sensing sensors detect these spectral signatures, allowing for the identification of salinity-affected areas and the assessment of salinity severity. Salinity mapping helps farmers implement targeted soil management practices, such as drainage, leaching, and salt-tolerant crop selection, to mitigate the impacts of soil salinization on crop productivity.

☆ **Integration with Environmental Data:** Remote sensing data can be integrated with environmental data, such as weather records, soil properties, and topographic information, to enhance abiotic stress monitoring capabilities. Geographic information systems (GIS) and machine learning algorithms enable the integration and analysis of multi-dimensional datasets for spatially explicit assessment of stress factors and their interactions. This integration provides valuable insights into the drivers and spatial variability of abiotic stress conditions, guiding adaptive management strategies for improving crop resilience and productivity.

Remote sensing technology provides a powerful tool for monitoring abiotic stress factors in agriculture, enabling early detection, spatial mapping, and informed decision-making to mitigate stress impacts on crop yields and ensure sustainable food production. By leveraging spectral, thermal, and environmental data, remote sensing contributes to proactive management of abiotic stressors, enhancing the resilience and productivity of agricultural systems in the face of changing environmental conditions.

Soil Moisture

Due to the advanced technology of Multispectral Photography and FTIR Spectroscopy, we are also able to detect the amount of moisture content present in a soil and water table. Soil moisture plays a critical role in agricultural productivity, influencing crop growth, water availability, and nutrient uptake. Remote sensing technology plays a significant role in monitoring soil moisture across large spatial scales and temporal frequencies. Here's how remote sensing contributes to soil moisture monitoring:

☆ **Active and Passive Remote Sensing:** Remote sensing sensors use both active and passive methods to measure soil moisture. Active sensors, such as radar and microwave radiometers, emit electromagnetic waves and measure their reflections from the Earth's surface. Passive sensors, on the other hand, detect naturally emitted radiation from the Earth, including thermal infrared and microwave radiation. Both active and passive remote sensing techniques provide valuable information on soil moisture content based on the interactions between electromagnetic waves and soil properties.

☆ **Microwave Sensing:** Microwave remote sensing is particularly effective for soil moisture monitoring because microwave radiation can penetrate through vegetation and cloud cover, allowing for all-weather and day-night monitoring capabilities. Microwave sensors measure the backscattered radiation or emission from the soil surface, which is influenced by soil moisture content. By analyzing microwave signals at different frequencies and polarizations, remote sensing can estimate soil moisture profiles at various depths and spatial resolutions.

☆ **Vegetation Indices:** Remote sensing data, such as multispectral imagery, can indirectly infer soil moisture conditions based on vegetation responses to water stress. Vegetation indices, such as NDVI (Normalized Difference Vegetation Index) and NDWI (Normalized Difference Water Index), are sensitive to changes in canopy water content and vigor. By analyzing changes in vegetation indices over time, remote sensing can provide insights into soil moisture availability and drought stress levels affecting crops.

☆ **Integration with Hydrological Models:** Remote sensing data are often integrated with hydrological models to estimate soil moisture dynamics and water balance components at regional scales. By assimilating satellite-derived soil moisture data into hydrological models, researchers can improve the accuracy of soil moisture predictions and enhance our understanding of water fluxes within watersheds. This integration enables timely assessments of water availability for agricultural irrigation, flood forecasting, and water resource management.

☆ **Precision Agriculture Applications:** Remote sensing technology is increasingly used in precision agriculture applications to optimize irrigation scheduling and water management practices. By monitoring soil moisture variability within fields using satellite or drone imagery, farmers can implement site-specific irrigation strategies based on actual crop water needs and soil moisture levels. This targeted approach minimizes water wastage, reduces irrigation costs, and maximizes crop yields while conserving water resources.

Remote sensing technology plays a critical role in soil moisture monitoring by providing timely, spatially explicit, and cost-effective information on soil moisture conditions across agricultural landscapes. By leveraging microwave sensing, vegetation indices, and integration with hydrological models, remote sensing enables informed decision-making in water resource management, irrigation planning, and agricultural sustainability efforts.

Crop Stands

Remote sensing helps in the identification and establishment of good crop stand to get maximum and uniform seedling and production. Remote sensing technology plays a significant role in assessing crop stands, which refer to the spatial distribution, density, and uniformity of crops within agricultural fields. Monitoring crop stands is essential for optimizing planting practices, assessing crop establishment, and identifying areas of potential yield variability. Here's how remote sensing contributes to evaluating crop stands:

☆ **Aerial and Satellite Imagery:** Remote sensing platforms, such as satellites and drones, capture high-resolution imagery of agricultural fields, providing detailed spatial information on crop stands. These images allow farmers and researchers to visually assess crop density, spacing, and uniformity across large areas. By analyzing aerial or satellite imagery, crop stands can be evaluated quantitatively using metrics such as plant count, canopy cover, and spatial distribution indices.

☆ **Vegetation Indices:** Remote sensing data, including multispectral and hyperspectral imagery, are utilized to derive vegetation indices sensitive to crop biomass and vigor. Indices such as NDVI (Normalized Difference Vegetation Index) and NDRE (Normalized Difference Red Edge) provide insights into crop health and density based on the differential reflectance of healthy vegetation compared to bare soil or non-vegetated areas. By analyzing changes in vegetation indices over time, farmers can assess crop stands' development and identify areas requiring additional management interventions.

☆ **Plant Health Monitoring:** Remote sensing technology enables the detection of stress factors affecting crop stands, such as nutrient deficiencies, pest infestations, or water stress. High-resolution imagery and spectral analysis techniques allow for the identification of subtle changes in crop reflectance associated with stress symptoms. By monitoring plant health indicators derived from remote sensing data, farmers can detect early signs of stress and implement targeted remedial actions to maintain optimal crop stands.

☆ **Spatial Analysis:** Remote sensing facilitates spatial analysis of crop stands' variability within fields, enabling farmers to identify areas of uniformity and heterogeneity. Geographic information systems (GIS) and spatial statistical techniques are employed to analyze crop stand data derived from remote sensing imagery, such as plant counts or canopy cover measurements. Spatial variability maps help farmers make informed decisions regarding variable-rate input application, precision planting, and site-specific management practices to optimize crop stands and maximize yields.

☆ **Temporal Monitoring:** Remote sensing allows for temporal monitoring of crop stands' evolution throughout the growing season. By capturing imagery at multiple time points, farmers can track changes in crop density, growth stages, and spatial distribution over time. Temporal monitoring enables the assessment of crop establishment success, emergence rates, and growth trajectories, facilitating timely interventions to address any deviations from desired crop stand conditions.

Remote sensing technology provides valuable insights into crop stands' characteristics and dynamics, empowering farmers with actionable information for optimizing planting decisions, assessing crop health, and maximizing yields in agricultural fields. By leveraging aerial and satellite imagery, vegetation indices, spatial analysis, and temporal monitoring, remote sensing enhances crop stand evaluation capabilities and supports data-driven management practices in modern agriculture.

Crop Canopy Measurement

Remote sensing also makes us easy to measure and maintain the canopy of a crop. So that we can apply an accurate amount of fertilizer, pesticide and any other chemical. Crop canopy measurement is essential for understanding vegetation structure, biomass accumulation, and overall crop health, which are critical factors in assessing crop growth and yield potential. Remote sensing technology plays a vital role in providing non-destructive and efficient methods for measuring crop canopy characteristics. Here's how remote sensing contributes to crop canopy measurement:

☆ **Vegetation Indices:** Remote sensing sensors capture spectral information from crops, which can be used to derive vegetation indices sensitive to canopy characteristics. Indices such as NDVI (Normalized Difference Vegetation Index) and NDRE (Normalized Difference Red Edge) are commonly used to quantify vegetation density, biomass, and vigor based on the differential reflectance of healthy vegetation across specific wavelengths. By analyzing vegetation indices derived from remote sensing data, researchers can estimate canopy cover, leaf area index (LAI), and biomass accumulation, providing insights into crop growth and development.

☆ **Lidar and Laser Scanning:** Remote sensing platforms equipped with lidar (Light Detection and Ranging) or laser scanning technology offer high-resolution, three-dimensional measurements of crop canopies. Lidar systems emit laser pulses and measure their reflections from the crop canopy, allowing for precise characterization of canopy height, structure, and volume. Laser scanning data enable the generation of detailed canopy profiles and digital surface models (DSMs), which facilitate accurate estimation of canopy parameters, such as canopy height, density, and architecture.

☆ **Hyperspectral Imaging:** Hyperspectral remote sensing sensors capture spectral information across hundreds of narrow spectral bands, enabling detailed characterization of crop canopy properties. Hyperspectral data provide spectral signatures for different crop components, such as leaves, stems, and soil background, allowing for discrimination and quantification of canopy characteristics. By analyzing hyperspectral imagery, researchers can assess biochemical composition, water content, and physiological status of crop canopies, facilitating precise measurements of crop health and stress.

☆ **Thermal Imaging:** Thermal infrared sensors mounted on remote sensing platforms measure the temperature emitted by crop canopies, which is influenced by canopy properties such as biomass, water content, and transpiration rates. Thermal imaging enables the estimation of canopy temperature differentials, which serve as indicators of plant water stress, disease presence, and overall canopy health. By analyzing thermal imagery, researchers can assess crop water status, detect stress conditions, and quantify canopy transpiration rates, providing valuable insights into crop water use efficiency and irrigation management.

☆ **Integration with Ground-based Measurements:** Remote sensing data are often integrated with ground-based measurements and field observations to validate and calibrate canopy measurement techniques. Ground-truth data, such as field surveys, canopy measurements, and biomass sampling, provide reference information for validating remote sensing-derived canopy parameters. Integration with ground-based measurements improves the accuracy and reliability of remote sensing-based canopy measurement methods, ensuring their applicability in agricultural research, crop management, and yield estimation.

Remote sensing technology offers versatile and efficient tools for measuring crop canopies and characterizing their structural, spectral, and thermal properties. By leveraging vegetation indices, lidar scanning, hyperspectral imaging, thermal sensing, and integration with ground-based measurements, remote sensing enables comprehensive and accurate assessment of crop canopy characteristics, contributing to improved understanding of crop growth processes, optimized management practices, and enhanced agricultural productivity.

Soil Property Sensing

To know the qualities and deficiencies of soil moisture, soil nutrients soil texture and soil structure. Remote sensing technology plays a crucial role in sensing and mapping soil properties across large spatial scales, providing valuable information for soil management, precision agriculture, and environmental monitoring. Here's how remote sensing contributes to soil property sensing:

☆ **Reflectance Spectroscopy:** Remote sensing sensors measure the reflectance of electromagnetic radiation from the Earth's surface, which varies based on soil properties such as texture, organic matter content, moisture levels, and mineral composition. Reflectance spectroscopy techniques analyze the spectral signatures of soil reflectance across different wavelengths, allowing for the estimation of soil properties through empirical relationships or statistical models. By correlating spectral reflectance with ground-based measurements of soil properties, remote sensing can provide spatially explicit maps of soil attributes, such as soil organic carbon content, texture classes, and moisture levels.

☆ **Hyperspectral Imaging:** Hyperspectral remote sensing sensors capture spectral information across hundreds of narrow spectral bands, enabling detailed characterization of soil properties. Hyperspectral data offer greater spectral resolution than traditional multispectral imagery, allowing for discrimination of subtle spectral features associated with soil composition and properties. By analyzing hyperspectral imagery, researchers can identify spectral signatures indicative of soil properties, such as clay content, soil moisture, organic matter, and mineral composition, facilitating accurate mapping of soil characteristics at fine spatial resolutions.

☆ **Soil Moisture Monitoring:** Remote sensing technology provides valuable information on soil moisture dynamics across agricultural landscapes. Microwave sensors mounted on satellites or aircraft measure microwave radiation emitted or scattered by the soil surface, which is influenced by soil moisture content. By analyzing microwave signals at different frequencies and polarizations, remote sensing can estimate soil moisture profiles at various depths and spatial resolutions. Soil moisture monitoring using remote sensing data supports irrigation management, drought monitoring, and water resource management in agriculture.

☆ **Thermal Imaging:** Thermal infrared sensors mounted on remote sensing platforms measure the temperature emitted by the soil surface, which is influenced by soil moisture levels, texture, and thermal properties. Thermal imaging enables the estimation of soil moisture status and thermal inertia, which serve as indicators of soil water content, soil texture, and soil heat flux. By analyzing thermal imagery, researchers can assess soil moisture distribution, detect anomalies associated with soil drainage or irrigation practices, and optimize water management strategies in agricultural systems.

☆ **Integration with Geospatial Technologies:** Remote sensing data are often integrated with geographic information systems (GIS) and other geospatial technologies to facilitate soil property mapping and analysis. GIS-based tools enable the spatial organization, visualization, and interpretation of remote sensing-derived soil data in conjunction with other spatial layers, such as land cover, topography, and climate data. This integration enhances the scalability and applicability of remote sensing-based soil property sensing, supporting soil mapping, precision agriculture, land management, and environmental monitoring initiatives.

Remote sensing technology offers versatile and efficient tools for sensing and mapping soil properties across agricultural landscapes. By leveraging reflectance spectroscopy, hyperspectral imaging, microwave sensing, thermal imaging, and integration with geospatial technologies, remote sensing enables comprehensive and accurate assessment of soil characteristics, supporting informed decision-making in soil management, agricultural practices, and environmental conservation efforts.

Geographic Information System (GIS)

Geographic Information System (GIS) is a comprehensive suite of computer tools used to acquire, store, retrieve, manipulate, and visualize geographical data from the physical world, serving a specific function or providing specific information. The process entails the acquisition, compilation, manipulation, organization, and representation of data within a digital context. Geographic Information System (GIS) is a robust collection of computer tools used to gather, store, retrieve, manipulate, and present real-world spatial data for specific purposes or information. The process encompasses data acquisition, data compilation, data manipulation, data modeling, and data visualization within a digital setting.

☆ Crop mapping and yield estimation.

☆ Erosion identification and remediation.

Applications of GIS in Horticulture

Geographic Information Systems (GIS) have emerged as indispensable tools in the field of horticulture, revolutionizing the way growers, researchers, and policymakers interact with spatial data. By integrating geographical information with powerful analytical tools, GIS offers a comprehensive platform for managing, analyzing, and visualizing spatial data related to horticultural resources, practices, and environments. From site selection and precision farming to market analysis and supply chain management, GIS facilitates informed decision-making, enhances productivity, and promotes sustainability across various facets of horticultural production and management. In this era of data-driven agriculture, the applications of GIS in horticulture continue to expand, driving innovation and efficiency in the cultivation, management, and distribution of horticultural crops worldwide.

Geographic Information Systems (GIS) play a crucial role in various aspects of horticulture, offering powerful tools for spatial analysis, decision-making, and management of horticultural resources. Here are some key applications of GIS in horticulture:

☆ **Site Selection and Land Suitability Analysis:** GIS enables horticulturists to identify suitable sites for establishing orchards, vineyards, nurseries, or other horticultural enterprises. By integrating spatial data layers such as soil types, topography, climate, and land use, GIS facilitates land suitability analysis to determine the most suitable locations for specific crops based on their environmental requirements and growth characteristics.

☆ **Crop Planning and Management:** GIS supports crop planning and management by providing spatial information on crop distribution, acreage, and planting patterns. Horticulturists can use GIS to optimize crop rotations, design planting layouts, and manage crop diversity within agricultural landscapes. GIS-based decision support systems help farmers make informed decisions regarding crop selection, spacing, irrigation scheduling, and pest management practices to maximize yields and profitability.

☆ **Precision Agriculture**: GIS plays a key role in precision agriculture applications in horticulture, enabling site-specific management of inputs and resources. By integrating remote sensing data, soil maps, and crop growth models, GIS helps farmers implement variable-rate fertilization, irrigation, and pesticide application strategies tailored to the specific needs of individual crops or crop zones within fields. Precision agriculture practices supported by GIS enhance resource efficiency, minimize environmental impacts, and improve crop productivity.

☆ **Pest and Disease Management:** GIS facilitates spatial analysis of pest and disease outbreaks in horticultural crops, enabling early detection, monitoring, and management of pest infestations and disease epidemics. By overlaying pest distribution maps with environmental variables such as temperature, humidity, and vegetation indices, GIS helps predict pest movement patterns and assess the risk of crop damage. Integrated pest management (IPM) strategies informed by GIS data promote sustainable pest control practices and reduce reliance on chemical pesticides.

☆ **Water Resource Management:** GIS supports water resource management in horticulture by analyzing spatial data on water availability, irrigation infrastructure, and crop water requirements. GIS-based hydrological models simulate water flow, infiltration, and runoff patterns within agricultural watersheds, helping farmers optimize irrigation scheduling, water allocation, and drainage systems. GIS also facilitates monitoring of water quality, groundwater recharge, and soil moisture levels, contributing to efficient water use and conservation efforts in horticultural production systems.

☆ **Market Analysis and Supply Chain Management:** GIS aids in market analysis and supply chain management by integrating spatial data on market demographics, transportation networks, and distribution channels. GIS-based market mapping helps horticultural enterprises identify target markets, assess demand trends, and optimize distribution routes for fresh produce. GIS also supports supply chain traceability and quality assurance by tracking the origin, handling, and distribution of horticultural products from farm to market, ensuring food safety and compliance with regulatory standards.

GIS enhances decision-making and resource management in horticulture by providing spatially explicit information and analytical tools for site selection, crop planning, precision agriculture, pest management, water resource management, market analysis, and supply chain management. By integrating GIS into horticultural practices, farmers, researchers, and industry stakeholders can improve productivity, profitability, and sustainability in horticultural production systems.

Global Positioning System (GPS)

GPS is a network of satellites, they continuously transmit coded information, which makes it possible precisely identify location on earth by measuring distancing from a satellite. Determination of position by satellite in digital form. The Global Positioning System (GPS) is a satellite-based navigation system that provides location and time information to users anywhere on or near the Earth's surface. Developed and operated by the United States Department of Defense, GPS consists of a constellation of satellites orbiting the Earth, ground stations, and user equipment.

- ☆ Relatively low-cost system, with no user charges.
- ☆ Available to users anywhere on the globe.

Applications of Global Positioning System in Horticulture

Global Positioning System (GPS) technology has become increasingly integral to modern horticulture, offering a myriad of applications that enhance efficiency, precision, and productivity in the cultivation and management of horticultural crops. By providing accurate positioning and navigation capabilities, GPS empowers horticulturists to optimize field operations, manage resources effectively, and make data-driven decisions tailored to the specific needs of individual crops and agricultural landscapes. From precision farming and site-specific management to navigation and mapping, GPS serves as a foundational technology that revolutionizes the way horticultural practices are planned, executed, and monitored, contributing to sustainable and profitable horticultural production systems globally. Global Positioning System (GPS) technology has numerous applications in horticulture, providing valuable tools for precision agriculture, navigation, mapping, and data collection. Here are some key applications of GPS in horticulture:

- ☆ **Precision Farming:** GPS enables precision farming techniques in horticulture by providing accurate positioning and navigation capabilities for agricultural machinery and equipment. GPS-guided tractors, sprayers, and harvesters allow farmers to perform precise field operations, such as planting, fertilizing, spraying, and harvesting, while minimizing overlaps and optimizing input use. Precision farming practices supported by GPS technology improve efficiency, reduce input costs, and enhance crop yields in horticultural production systems.

- ☆ **Site-Specific Management:** GPS facilitates site-specific management of horticultural crops by georeferencing field boundaries, soil sampling points, and management zones within agricultural landscapes. GPS-based soil mapping and sampling techniques enable farmers to assess soil variability, nutrient status, and pH levels across fields, guiding targeted soil amendments and fertilization strategies tailored to the specific needs of individual crop zones. Site-specific management practices supported by GPS enhance resource efficiency and optimize crop production in horticulture.

☆ **Navigation and Guidance:** GPS aids in navigation and guidance for horticultural operations, such as orchard management, vineyard maintenance, and nursery operations. GPS-enabled handheld devices and mobile applications provide real-time positioning information and navigation assistance to farmers and field workers, facilitating efficient routing, task assignment, and monitoring of field activities. GPS navigation systems improve operational efficiency, reduce labor costs, and enhance productivity in horticultural enterprises.

☆ **Mapping and Surveying:** GPS technology supports mapping and surveying applications in horticulture, enabling accurate delineation of field boundaries, crop rows, planting layouts, and irrigation infrastructure. GPS receivers integrated with Geographic Information Systems (GIS) allow for precise mapping of spatial data layers, such as soil types, topography, drainage patterns, and land use, which are essential for site selection, land management, and decision-making in horticultural production. GPS-based mapping and surveying tools enhance spatial analysis capabilities and support data-driven management practices in horticulture.

☆ **Data Collection and Analysis:** GPS facilitates data collection and analysis for monitoring crop performance, assessing field conditions, and evaluating horticultural practices. GPS-enabled sensors and data loggers record spatially referenced measurements of environmental variables, such as temperature, humidity, and light intensity, at multiple locations within agricultural fields. These data are used to generate spatial maps, time-series analyses, and statistical models for crop monitoring, yield estimation, and research purposes in horticulture. GPS-based data collection and analysis tools improve data accuracy, reliability, and efficiency in horticultural research and management.

GPS technology offers versatile applications in horticulture, enabling precision farming, site-specific management, navigation and guidance, mapping and surveying, and data collection and analysis. By leveraging GPS technology, horticulturists can enhance operational efficiency, optimize resource use, and improve crop productivity in diverse horticultural production systems.

Chapter 9

Precision Farming in Horticultural Crops

Introduction: An Overview

Precision farming is a highly scientific approach to achieving sustainable agriculture and horticulture. Precision farming has a lengthy historical background, commencing in the early 1900s and gathering momentum towards the conclusion of the century. However, there is a substantial amount of work yet to be accomplished. Horticulture precision farming involves the utilization of many technologies and methodologies to identify, assess, and manage variations within agricultural fields, with the

Precision farming in horticulture aims to achieve two goals: maximizing yield and minimizing environmental degradation. Precision technology, with its immense potential for economic expansion in the global market, is also characterized by exceptional quality. While extensively utilized for commercial crops in several industrialized nations and a handful of emerging nations, it remains in its early stages in the majority of developing countries. Remote sensing technology in horticulture can provide a crucial input in the form of a variability map for precision farming. Precision horticulture holds significant promise in poor nations, but its implementation necessitates substantial effort and integration. Several precision farming initiatives currently underway in India are anticipated to yield significant results and transform Indian horticulture from a subsistence-based activity to a profitable commercial enterprise. In India, where the majority of people rely on the agriculture and horticulture sectors, technologies such as precision farming and remote sensing can contribute to the enhancement of the socio-economic status of these farmers. This, in turn, can lead to income generation and overall development of the country.

aim of optimizing profitability, sustainability, and land resource conservation, while avoiding negative effects on the environment. Despite extensive study, only a minority of farmers have implemented precision technologies to restructure the agricultural system towards low-input, high-efficiency, and sustainable practices. It can be conceptualized as a comprehensive framework for enhancing horticulture production by utilizing crop-specific data, cutting-edge technology, and effective management strategies. It is referred to as a comprehensive system that encompasses crop or commodity planning and extends to the post-harvest processing stage of production. Success in a production system hinges on three important factors: information, technology, and management. Information on crops is an extremely useful asset for contemporary farmers. Accurate data is essential at all stages of the production process, starting from the initial planning phase and continuing through post-harvest management. Information required during the field production stage encompass spatial and temporal data pertaining to the soil, crop, pest, topography, and weather. Temperature, humidity, and moisture are crucial variables during the post-harvest phase. Additional information can be obtained from previous crop records. Additional data must be acquired in real-time to enable immediate utilization by the system. Technology is considered the system's second most crucial element. For precision farming to be successful, it is necessary to have production equipment and systems that are specifically designed to meet the operational requirements of precision farming. The advancement of accurate planting and chemical-application technology is recognized as the foundation for precision farming, from a mechanized perspective. Precision equipment enables accurate control and delivery of agricultural pesticides, allowing for variable-rate treatments. This foundation also includes the global positioning system (GPS), geographic information systems (GIS), and computers and the internet. GPS with differential correction has shown to be a useful tool for geo-referencing field descriptions and data. The ability to organize data by geo-referenced circumstance is provided by GIS. Use of computers and the internet for analysis and control in order to construct the comprehensive system required in site-specific and post-harvest management systems. Management is the third pillar of success. Management provides us with the tools we need to analyze data and make informed production decisions in a fast and efficient manner. Management practices can be used to attain precision farming objectives using the information and technology available to today's farmer. Information and technology contribute very little to the effectiveness of the production system if management practices are ineffective. Precision farming is a method of growing crops that uses exact amounts of inputs to produce higher average yields than traditional methods. One of the primary issues in India is the limited size of the field. In the country, about 60 per cent of farmers have operational holdings of less than one hectare (ha). Only 20 per cent of the states *i.e.* Punjab, Rajasthan, Haryana, and Gujarat have operating holding sizes greater than four hectares. Precision farming opportunities in cooperative farms are expanding in both business and horticulture crops. Precision farming, which is based on the use of technology, is the most valuable farm management invention of this century.

This technique is the latest advancement in sustainable horticulture and nutritious food production. It offers enhanced profitability, productivity, and economic efficiency, while also reducing negative environmental impacts. Emerging precision horticulture technologies heavily depend on remote sensing geographic information systems (GIS), global positioning systems (GPS), auto analyzers, sensors, computers, and suitable software and hardware to accurately identify areas of nutrient deficiencies and other biotic and abiotic stresses. At the local level, the economic importance of elements that limit soil-water-fertilizer-pest-crop interactions, as well as their environmental consequences. They can assist in the implementation of comprehensive management systems for nutrients, soil health, water, pests, energy, and diverse crop genetic resources. The main objective of implementing precision farming in India is to enhance horticulture production, improve environmental quality, and strengthen the economic situation of farmers.

In modern times, developed nations prioritize enhancing their output by utilizing precision farming technology. They are aware that land is finite and its expansion is limited. Therefore, they utilize technology to meet their food demands without increasing the land area or disrupting the ecological balance. Precision technology is essential for ensuring high quality while minimizing the use of land resources. The use of various technologies such as GPS, GIS, and remote sensing simplifies the application of fertilizers, pesticides, and other substances to specific locations where they are needed. These technologies identify the sites that require improvement and determine the precise requirements for water, pesticides, and fertilizers. Remote sensing technology is employed to determine the overall extent of cultivable land in a given location, including the portion that is unused or considered wasteland. Additionally, it helps identify areas where the soil poses challenges or difficulties. Government institutions utilize this technology to provide data on several aspects such as the overall size of the country, cultivated land, net sown area, and waste land. This data contributes to the enhancement of the horticulture and agriculture sectors, as well as the overall prosperity of the nation. Developed nations are continuously striving to optimize these technologies, but India has certain obstacles that hinder its adoption and optimization of precision technology. The limits include a shortage of educated and innovative farmers, limited land holdings, government laws, a lack of extension programs, and other factors. However, in order to meet the needs of a growing population in the future, it will be necessary to embrace advanced technology such as precision farming. Precision technologies are crucial for the future of agriculture and horticulture. The growth of a nation's production system is dependent on these technologies. The increasing demand for high-quality production poses a challenge for developing and undeveloped nations to compete in the food and commercial market.

Importance of Horticulture Precision Farming

In the present era, technology is revolutionizing various industries, including agriculture, as it continues to incorporate advanced technologies. Horticulture precision farming is a technology that offers significant advantages in terms of both quantity and quality, while minimizing its impact on the environment. Like other technologies, it also decreases the amount of time required and gathers more precise data through the use of GPS and GIS technologies. The necessity of this technology in the field of horticulture may be comprehended through the following points:

- ☆ To mitigate soil degradation.
- ☆ Optimal utilization of water resources.
- ☆ Cultivating positive attitudes.
- ☆ To enhance the efficiency of horticulture production.
- ☆ Creation of technical job opportunities.
- ☆ To preserve ecological equilibrium.
- ☆ Reduction of chemical usage in agricultural crop production. Precision farming technology is altering the socio-economic standing of farmers.
- ☆ To contribute to the attainment of the objective of sustainable agriculture/ horticulture.
- ☆ Contemporary methodologies for enhancing production quality, quantity, and cost reduction.

Basic Steps in Horticulture Precision Farming

It is based on field variations such as crop qualities, soil qualities, and so on. These characteristics or variances are often recorded and mapped. Assessing, controlling, and evaluating variability are the main stages that contribute to the framework of horticulture precision farming.

- ☆ **Assessing Variability:** Evaluating the range of differences is the initial stage in achieving accuracy in horticulture farming. It is a crucial stage because it is evident that one cannot effectively handle anything that one does not comprehend. The primary reason of variation in these processes is predominantly attributed to geographic and temporal variability, but the simultaneous reporting of both variables is infrequent. Surveying, interpolating point samples, utilizing high resolution aerial and satellite data, and employing modeling techniques are all strategies that can be utilized to map spatial variability in the field.

- ☆ **Managing Variability:** After thoroughly analyzing the variation in the field, the farmer should adjust their crop and agronomic inputs to fit the individual conditions using site-specific management recommendations and the right equipment to control it. The effectiveness of precision farming relies on the farmers' capacity to control insect infestations, soil fertility, crop management, water management, stress management, and other factors. It also depends on how accurately they implement corrective actions in the field to address detected variations. These traits can be documented and mapped for effective precision farming systems.

☆ **Evaluation of Precision Farming:** The evaluation of precision farming relies on three essential factors: economic variability, environmental conservation, and the potential for transferring precision agricultural technologies. The first aspect is economic evaluation, which examines whether the reported agricultural benefits are being effectively transformed into economic value through market mechanisms. The second study focuses on the environmental impact of precision farming and evaluates its potential to enhance the sustainability of land, water, and the overall environment in agricultural systems. The third and most essential inquiry is to the viability of implementing site-specific technological farming on individual farms, as well as the extent to which this cutting-edge technology can be shared with other farmers.

Horticulture Precision Farming: Tools and Technologies

Precision farming in horticulture functions as a highly coordinated system, integrating several elements to facilitate and streamline crop management. In

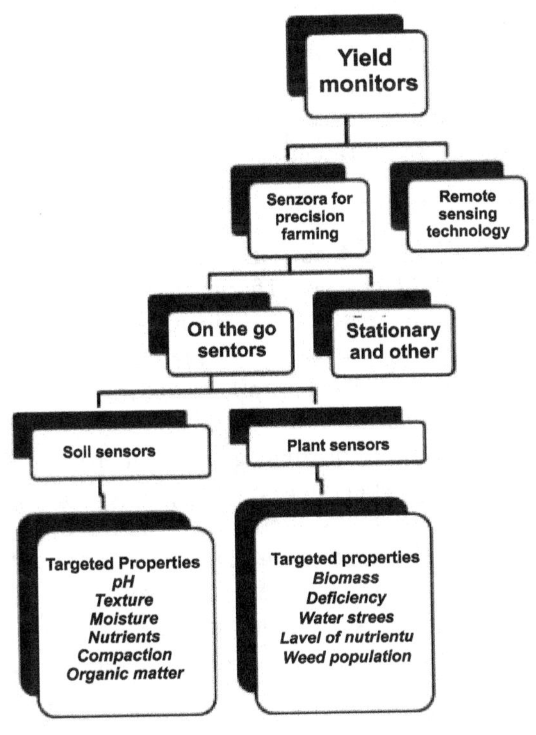

precision farming, each component has a crucial role, working together to offer the best possible crop care. The utilization of data sensors and automated machinery in horticulture has resulted in a significant transformation, leading to increased efficiency and sustainability. Incorporating these sophisticated technology and meticulousness serves as a solid foundation for the contemporary agricultural method, guaranteeing a perceptive and data-oriented approach to farming. Now, let's examine the fundamental elements that render precision farming a formidable asset for contemporary horticulturists and farmers.

Precision farming in horticulture represents a transformative approach to cultivation, blending traditional agricultural practices with cutting-edge technologies to optimize crop yields and resource efficiency. By leveraging advanced tools such as remote sensing, GPS guidance systems, and data analytics, precision farming enables growers to make informed decisions tailored to the specific needs of plants. This integration not only enhances productivity but also promotes sustainability by minimizing inputs such as water, fertilizers, and pesticides. As horticulture continues to evolve in the face of global challenges like climate change and food security, precision farming emerges as a pivotal strategy in shaping the future of agricultural practices worldwide.

Computer and Internet Technology

Computers and Internet technologies are crucial tools for enabling any form of technology, including precision farming. They serve as the main means of collecting, processing, and gathering data. The utilization of high-speed computer systems has significantly contributed to the advancement of technology by enabling faster processing, efficient data collection, and the execution of highly accurate activities. Its enhanced accuracy, together with a decrease in time, makes it a perfect element for incorporation into any form of technology or horticultural technology.

Global Positioning System (GPS)

Yield mapping and variable rate fertilizer and pesticide application are the most common uses of GPS. The GPS is essential for accurately detecting the exact location in the field in order to monitor spatial variability and apply inputs particular to each site. When operating in differential mode, GPS may achieve a location accuracy of 1 meter. In the future, farmers will have the ability to carry out farming operations at any given moment due to the utilization of high-precision GPS technology. Optimal conditions for spraying may occur during nighttime when wind speeds are minimal. To reduce the germination of specific weeds triggered by light, employ nocturnal tillage. From an agricultural standpoint, the GPS system should fulfill the following requirements: consistent performance in diverse terrains, updates of position at a minimum rate of one per second, ability to map crop yield using a combine with a cutting width of 5 meters, location accuracy within a range of +3 meters, and precision of 10 meters for applications based on changes in soil type.

Geographical Information System (GIS)

The geographic information system (GIS) serves as the central component of precision farming. A highly structured system comprising computer software, hardware, geographic data, and personnel that efficiently captures, stores, updates, manipulates, analyzes, and presents various types of geospatially connected information. The spatial analytical capability of GIS technology is commonly known as GIS technology's spatial analytical capability. It can assist with horticulture in two distinct manners. An approach involves the linkage and integration of Geographic Information System (GIS) data with simulation models, including those pertaining to crop, soil, and weather field records. The second purpose is to support the engineering department in creating precision agriculture tools and GPS-guided machinery, namely a variable rate applicator. By integrating soil, vegetation, and meteorological data, it is possible to ascertain the potential yield of a field, on the assumption that no other factors disrupt the typical growth of plants. With the utilization of this method, we can promptly detect issues in the vicinity, ascertain the reason for the decrease in field strength, and thereafter implement suitable actions to rectify it.

Remote Sensing Technology

Remote sensing is a method of precision farming that uses satellite-based sensors or CIR video digital cameras mounted on small planes to collect images. Multiple institutions are employing this remote sensing technology due to its significant capacity for monitoring temporal and geographical changes over time with high precision, making it well-suited for precision farming.

Variable Rate Technology (VRT)

The current ground machinery equipped with ECU and GPS technology is capable of meeting the variable rate input demand. This method utilizes optimal input rates, minimizes expenses, and maintains a harmonious balance with the environment without compromising production and quality. The VR system typically employs two methods: map-based and name sensor-based. A map-based system often relies on current and historical data. The sensor-based technology enables the gathering of information and the adjustment of fertilizer rate, pesticide rate, and seed rate for application on the field. It detects the soil and crop characteristics and adjusts the application equipment accordingly. The Farmers are more receptive to this system.

Yield Mapping Technology

Yield is the most precise measure of variation in different cultivation parameters throughout a field. Hence, the process of mapping and analyzing crop yields, along with interpreting and correlating them with spatial and temporal variations of other indicators, plays a crucial role in formulating crop management strategies for the upcoming season. Modern yield monitors utilize advanced technology to quantify the rate of volume or mass flow, enabling them to accurately record the amount of harvested crop during a specific time period. A comprehensive color-coded thematic map is generated by syncing time-based yield data with location addresses obtained from the onboard GPS device. Implementing yield mapping in mechanized crops is straightforward. In order to determine the quantity and mass of fruit collected, two methods were employed: either by utilizing load cells positioned on a conveyor belt to weigh the crop as it passed through, or by employing an arrangement of sonic beams put above the grape discharge chute. The findings showed a significant difference in yield, ranging from 8 to 10 times, between different sections of the same plot of land. Research on arable crops has demonstrated that by the third year, the patterns of yield fluctuation become nullified, resulting in the emergence of consistent areas with high and low yields, as well as unpredictable production variations.

Data Sensors

Data sensors serve as the watchful observers of precision farming in horticulture, gathering up-to-the-minute data on soil conditions, crop health, and environmental factors. These sensors offer a continuous flow of data, enabling farmers to make well-informed decisions. Horticulturists can customize their interventions by monitoring these insights, guaranteeing that each plant receives the exact care it needs for ideal growth.

Automated Machinery

Automated technology is the cornerstone of precision farming, carrying out operations with unmatched precision and efficiency. These technologies, ranging from autonomous planters to harvesters, perform tasks using insights derived from data. This not only decreases the amount of work required but also guarantees that activities such as planting, harvesting, and spraying are carried out accurately, resulting in less waste and more output in horticulture.

Drones for Aerial Insights

Drones elevate precision farming to unprecedented levels. These autonomous aerial vehicles acquire high-resolution photographs and data, offering a top-down perspective of fields. By utilizing an airborne perspective, farmers may effectively discern crop variances, evaluate plant vitality, and spot potential problems such as pest infestations. Drones provide a rapid and thorough assessment of crops and are the most effective method for monitoring extensive areas, allowing for timely and focused treatments.

Data-driven Decision Support Systems

Decision support systems function as the intellectual capacity behind precision farming, analyzing the large quantities of data gathered by sensors and other technology. These systems utilize data analysis to offer practical and valuable insights, assisting farmers in making well-informed decisions. Through the utilization of data analytics, horticulturists and farmers can make informed decisions on the allocation of resources, mitigate the risks associated with crop decay or failure, and implement precise interventions to improve crop management.

Potential of Precision Farming in India

Precision farming is a proven sophisticated technology that has been successfully implemented in many industrialized countries. It has the ability to meet the demanding needs of any country, provided it is applied correctly and with the right technological approach. Precision farming in India has had limited reach thus far. Nevertheless, there are other horticultural crops in India that have a lucrative market and have ample opportunities for precision farming, but this potential has not yet been acknowledged. The practice of site-specific farming in India is restricted by several restrictions. The constraints in India's agricultural sector include limited land holdings, imperfect markets, insufficient local technical expertise, knowledge and technological gaps, low levels of education and awareness, variations in cropping systems, the socio-economic status of farmers, limited availability of high-quality and cost-effective data, inadequate cost-benefit analysis of precision farming, and a shortage of entrepreneurs in the field. Components of Precision Farming is depicted in Figure 9.1.

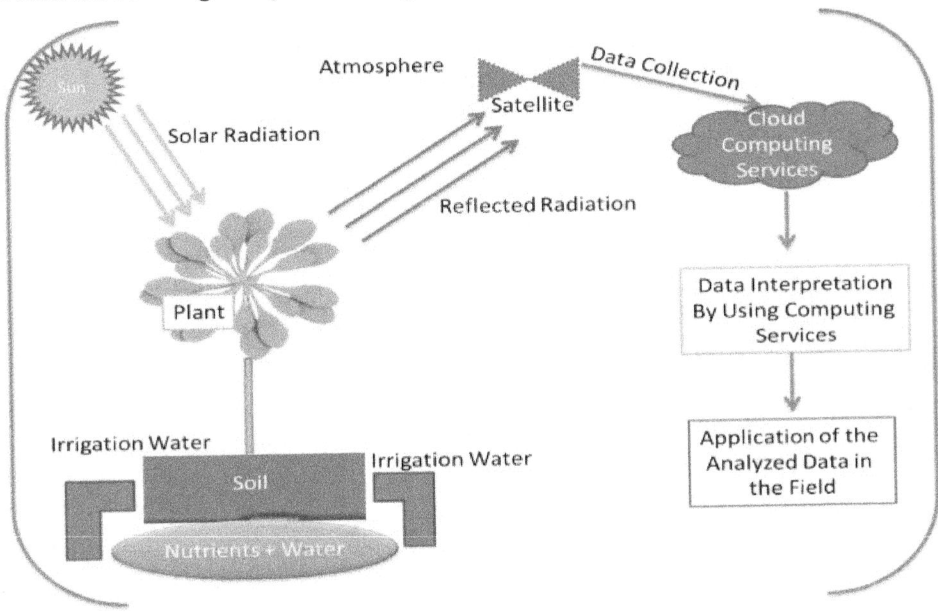

Figure 9.1. Components of Precision Farming.

Special Applications of Precision Farming in Horticulture

Horticulture greatly relies on precision farming for optimal results. This is an innovative method used in agriculture to provide customized solutions for horticulture farmers. It aims to improve the effectiveness and sustainability of their horticultural activities. The specific applications of this technology result in a significant transformation in the methods used to maintain and develop crops:

- ☆ **Precision Irrigation:** Precision irrigation involves the use of sensors and monitoring systems by horticulture growers to assess soil moisture levels and provide water precisely to the areas and times where it is required. This practice conserves water, mitigates waterlogging, and safeguards against illnesses in plant roots.

- ☆ **Fertilizer and Pesticide Management:** These farmers utilize technology to assess the soil and crops in real-time, enabling them to give the exact amount of nutrients and pesticides needed by the crops. This conserves resources and minimizes the environmental footprint.

- ☆ **Disease and Pest Detection:** Precision farming allows farmers to utilize automated devices equipped with sensors to promptly detect diseases and pests. By selectively targeting impacted regions, farmers can reduce the use of detrimental pesticides and safeguard the health of crops and the environment.

- ☆ **Yield Prediction and Harvest Optimization:** Horticulture farmers employ precision farming equipment to assess data on crop health, weather patterns, and soil conditions in order to make well-informed judgments on the optimal timing for harvesting, resulting in maximum output. This enhances the yield and caliber of the harvest.

- ☆ **Protected Cultivation (Greenhouses and Tunnels):** Protected cultivation, through the use of greenhouses and tunnels, allows farmers to construct controlled environments that provide optimal growing conditions for plants. This enables farmers to cultivate a diverse range of crops that are not native to their nation. This also enables farmers to plant crops in places with harsh temperatures, hence enhancing total agricultural output and quality.

Advantages of Precision Farming

- ☆ Precision farming provides easy management of arable land in large area and reduces the time.

- ☆ It provides technological support to produce more qualitative matter than the traditional system.

- ☆ To manage the non-uniform land through divided into smaller plots according to specific need.

- ☆ Optimization of agrochemical products.

- ☆ Optimization of water resources and technical expertise

- ☆ To provide chances for better resource use and reduce wastage.

- ☆ It minimizes the maximum risk to the environmental factors.

☆ Global positioning system technology allows surveying of the field.

☆ Mapping technology allows mapping of yield and soil characteristics

Disadvantages of Precision Farming

☆ Preliminary cost may be high.

☆ Need of technical expertise in these areas.

☆ Extremely demanding effort predominantly collecting and analyzing data.

☆ It should be seen as long-term investment.

☆ It may take some years to fully implement the system.

Application of Precision Technologies in Horticultural Crops

☆ The banana is one of the crops in India that has benefited greatly from precision farming techniques. Micro propagation, fertigation, crop geometry, drip irrigation, green manuring, mulching, recycling of banana waste, organic farming, proper hygiene of banana plantations through integrated disease and pest management, training, processing to puree transfer technology, participatory demonstration, and so on are some of the factors that contribute to the success of the banana industry.

☆ Papaya, a tropical fruit, is extensively cultivated, with India being the leading global producer. Propagation and nursery production, variety selection, planting season, sex expression, spacing, thinning, intercrop, irrigation management, plant protection, and harvesting are all specific techniques that can improve yield.

☆ The Aonla is native to the Indian subcontinent. The precision system for improved production encompasses various components such as multiplication of authentic planting material, adoption of effective planting techniques, careful selection of varieties, proper training and pruning methods, appropriate application of manure and fertilizers, efficient water management, effective pest and disease control, as well as careful harvesting and post-harvest management.

☆ Guava is considered to be one of the most nutritionally beneficial and financially lucrative crops. Guava fruits are used for both fresh and processed consumption. Precision farming involves the careful selection of types, the propagation of authentic planting material, and the creation of a guava orchard, all aimed at increasing productivity. During the establishment process, various factors are taken into account, including planting techniques, training and pruning, high density planting, rejuvenation of old plantations, growth and development, weed control, irrigation, inter cultivation, crop regulation, nutrition, fertilisation, and effective management of deficiencies, pests and diseases. Harvesting is also an important consideration.

Post-harvest Management through Horticulture Precision Farming

In India, the post-harvest management of horticulture crops is significantly inferior compared to other countries, resulting in substantial losses in both quantity and quality. These losses have a direct impact on the income and socio-economic status of the farmers. Post-harvest management commences following the completion of agricultural harvesting. At this stage, inadequate handling can have a significant negative impact on the quality. The application utilizes sensors to monitor the conditions during the curing or storage process in order to meet the objective of achieving optimal parameters and quality of the preserved or edible material. Manual monitoring is not feasible due to its labor-intensive nature. However, sensor technology enables efficient handling of the monitoring process. The quality and income of certain horticultural crops can be improved by continuously monitoring them in cold storage and adjusting their conditions based on demand.

Benefits of Precision Farming in Horticulture

Precision technology in horticulture is an emerging and rapidly advancing technology that employs sophisticated techniques and data to revolutionize plant cultivation and maintenance. These technologies collaborate to deliver a more effective, environmentally friendly, and accurate method for managing crops. They serve as a comprehensive set of tools for farmers, encompassing sensors, GPS technology, automated machinery, drones, and decision support systems. Sensors serve as the visual and auditory organs of the farm, gathering up-to-the-minute information on soil moisture, temperature, and crop well-being. This data assists horticulturists and farmers in making well-informed decisions and ensuring that each plant receives the appropriate quantity of water, nutrients, and attention to promote robust growth. GPS technology enables farmers to accurately monitor their fields, facilitating exact mapping and targeted engagement in activities such as planting and harvesting. Automated machinery might be considered the workforce of precision horticulture. It performs duties with precision and efficiency, relieving farmers of physical exertion and enhancing productivity. In addition, drones offer an aerial perspective of fields, gathering detailed photographs and data. Decision support systems then analyze the data acquired from different technologies to provide practical insights, enabling farmers to make well-informed decisions. As we examine the potential benefits of incorporating precise technology into farm operations, it is crucial to identify the specific areas in which this technology might assist horticulturists and farmers. Although precision farming is a recent technology, its primary objective is to establish sustainable and conscientious farms that minimize waste and prioritize resource optimization. Here is an examination of the advantages that precision farming brings to the field of horticulture.

☆ **Resource Efficiency:** Resource efficiency is achieved by precision farming, which enables precise management of crucial components such as water, fertilizers, and pesticides. By employing a focused strategy, the allocation of resources is optimized, guaranteeing that every plant has the precise amount of nutrients necessary for optimal development.

- ☆ **Increased Crop Produce:** Enhanced Agricultural Output: Precision farming techniques, through the optimization and refinement of cultivation methods, result in increased agricultural productivity. Farmers' newfound capacity to cater to the distinct needs of each plants, encompassing factors such as soil composition and microclimate fluctuations, results in more robust and fruitful yields from the planted seeds.

- ☆ **Ensuring Sustainability:** Precision farming helps mitigate the environmental impact of horticulture. By understanding the exact needs of the plants, farmers can reduce the dependency on agrochemicals and optimise irrigation. It also helps promote sustainable use of available resources minimizing pollution and preserving natural resources.

- ☆ **Data-Driven Decision Making:** Precision farming enables farmers to utilize real-time data and analytics to make well-informed judgments. Access to precise information enhances the capacity to rapidly address difficulties, ranging from monitoring crop health to predicting potential threats.

- ☆ **Labor Efficiency:** By integrating automated machinery and robots, farmers may optimize the use of labour by reducing the need for manual chores, allowing human resources to be allocated to more strategic and sophisticated parts of farming. This not only enhances productivity but also tackles labour shortages in the agricultural industry.

- ☆ **Risk Reduction:** Precision farming facilitates risk management by enabling early identification of possible problems in crops, weather, and soil. Whether it involves detecting insect infestations on crops or predicting unfavourable weather conditions that could impact crop viability. This enables farmers to anticipate and implement proactive strategies to protect crops and limit damages.

- ☆ **Enhance Economic Viability:** Although precision farming requires an initial investment in technology, its long-term advantages significantly contribute to increasing the economic sustainability of horticulture. By leveraging their newfound comprehensive understanding, farmers can achieve increased crop output, lower expenses related to resources, and optimize their operations to ultimately enhance the overall profitability of their agricultural endeavours.

- ☆ **Adaptability to Climate Change:** Precision farming provides versatility in response to the problems posed by climate change. Farmers can enhance crop resilience in an unpredictable climate by utilizing real-time data to adapt cultivation operations according to changing climatic conditions.

Chapter 10

Precision Farming of Fruit Crops

Introduction

Precision farming, particularly in the cultivation of fruit crops, represents a paradigm shift in agricultural practices aimed at enhancing efficiency, productivity, and sustainability. Unlike traditional farming methods, which often rely on broad and uniform application of inputs such as water, fertilizers, and pesticides, precision farming leverages advanced technologies to tailor these inputs precisely to the needs of individual plants or small sections of fields. This approach not only optimizes resource use but also minimizes environmental impact and maximizes crop yield and quality. In the context of fruit crops, where factors like climate variability and pest pressures can significantly affect outcomes, precision farming offers a promising pathway to meeting the growing global demand for high-quality produce while ensuring long-term agricultural resilience.

Precision Farming in Banana Production

Precision Farming is a farm management approach that use information and technology to identify, analyze, and manage variability within fields. Its goal is to achieve maximum profitability, sustainability, and protection of the land resource. Utilizing this farming method, advanced information technologies can be employed to enhance decision-making processes about various elements of crop cultivation. Precision farming entails analyzing and addressing the inherent variability within a field to achieve enhanced efficiencies. The objective is not to achieve uniform yield across all areas, but rather to strategically allocate and distribute resources based on the individual characteristics of each site in order to optimize long-term cost-effectiveness. The cultivation of bananas holds a key position in India's fruit industry. Maharashtra has the top position in banana productivity, with a rate of 62.9 tones per hectare. This surpasses the national average of 33.5 tones per hectare. Additionally, Maharashtra is one of the few regions in India where precision farming techniques have not only been invented, but are also being effectively implemented, resulting in significant benefits. Enhancements in banana productivity can be attributed to several factors such as varietal development, micro propagation, crop geometry, drip irrigation, fertigation, bunch management, intercropping, mulching, propping, and integrated pest and disease control.

Varietal Improvement

Efforts in varietal improvement have been concentrated on the development of cultivars that exhibit high productivity and resilience against significant biological and environmental pressures. The Cavendish group had conducted several clonal selections. The main cultivars grown in Maharashtra include Basrai, Shrimanti, Ardhapuri, Mahalaxmi (Robusta), and Grand Naine, which are cultivated throughout the entire state of Maharashtra. These cultivars collectively account for almost 90 per cent of the total area. These cultivars are classified as dwarf to semi-dwarf varieties and have a crop period of 12-15 months, which varies according on the soil type, planting timing, and other management factors. These cultivars have a conical to cylindrical shape of bunch, with a weight ranging from 15 to 30 kg. Many of these cultivars have moderately to evenly spaced clusters of slightly to moderately curved fruits, which are green to dark green in color. The cultural procedures for the majority of these dwarf Cavendish cultivars are comparable.

Propagation Techniques

In vitro Techniques

In Maharashtra, the traditional method of propagating banana is by suckers. However, this method has led to the spread of illnesses and pests, which has made it necessary to use clean material for propagation. In this particular setting, the techniques of macro and micro propagation were created. Tissue-cultured plantlets have gained economic importance due to their consistent crop quality, early maturity, resistance to diseases and pests, and high productivity. A single growth point can yield approximately 2,000 plantlets by sub culture during a year. India requires 746 million plantlets to be produced using tissue culture in order to cover at least one third of the banana cultivation area. The manufacture of tissue culture plants began in 1990 with the production of 200,000 to 300,000 plants. By 2010-11, the production had increased to 120 million plants. The NRC for Banana in Trichy and NCL in Pune have established standardized tissue culture techniques for the successful in vitro propagation of a wide range of cultivars. An optimal tissue culture plant should have a hardening age of 45-60 days, a height of 30 cm, 5 to 7 leaves, and a pseudo stem circumference of 5.0-6.0 cm.

In vivo Techniques

The micropropagation technology efficiently produces a substantial quantity of banana planting material in a very brief timeframe. Nevertheless, the accessibility of tissue culture plants for farmers is restricted and comes with elevated expenses. These tissue culture plants are also displaying unfavourable deviations on the farm. The process of natural regeneration in Banana is hindered by the hormone-mediated apical dominance of the mother plant, resulting in delayed growth. Nevertheless, the inhibition of apical dominance to promote the growth of lateral buds and enhance the pace of suckering can be achieved using mechanical methods such as complete or partial decapitation, or by using the detached corm approach.

Planting System

The banana producing system has been standardized to incorporate high density planting. It is recommended to plant Cavendish group cultivars such as Basrai, Shrimanti, and Grand Naine in a high density square planting pattern at a spacing of 1.5 meters by 1.5 meters. The land can support a density of 4,444 plants per hectare. In this ecosystem, there is intense rivalry for sunlight. However, this competition helps to fully cover the canopy area and prevent radiation from reaching the soil. It alters the garden's microclimate by lowering the soil temperature, inhibiting weed development, reducing evaporation from the soil, and increasing the relative humidity. It also provides protection for the plant and clusters from intense heat during the summer months, when temperatures can reach up to 48 degrees Celsius, as well as from damage caused by strong winds. In Maharashtra, it is also recommended to use the pair row strategy of planting at a spacing of 0.9 X 1.5 X 2.1 meters. It may also accommodate 4,444 plants per hectare. Farmers in western Maharashtra are also embracing spacing options of 1.5 X 1.8 m, 1.8 X 1.8 m, or 1.8 X 2.1 m. The Banana Research Station in Jalgaon has advised that banana planting in Maharashtra state should be done in June-July (Mrig Baugh) and October (Kande Baugh). Approximately 70 per cent of planting in Maharashtra occurs in the months of June and July, while 20-25 per cent of planting takes place in October.

Protected Cultivation

In the 1980s, the practice of protected farming of bananas became increasingly popular, particularly in Morocco and the Canary Islands. These nations are located in the Mediterranean region and experience harsh winter temperatures. Between 1984 and 2009, the cultivation of bananas using plasticulture in Morocco has shown a significant growth, expanding from 5.0 hectares to 4,000 hectares. Banana farming in protected environments, such as greenhouses, is becoming increasingly significant in Turkey. Currently, 2,500 hectares out of a total area of 4,000 hectares are dedicated to greenhouse cultivation. The study conducted by Galan Sauco *et al.* (1992) in the Canary Islands clearly showed that banana trees exhibited greater size, heavier bunches, shorter cycle intervals, and considerably improved annual production per hectare when grown under plastic cultivation compared to open cultivation. The impact of these factors was more noticeable on colder slopes that face north, where the limitations imposed by the climate are more extreme. The significant increase in crop production demonstrated that utilizing a plastic greenhouse was economically viable, with the potential to recoup its initial investment within a span of 5 years. The exorbitant expense of greenhouses is a significant limitation for covered farming. However, in Israel, as well as formerly in the Canary Islands, any new banana plantings that utilize net protection instead of open air cultivation are eligible for a 50 per cent subsidy to cover the cost difference (Y. Israeli, 2009, personal communication).

Nutrient Management

Naturally, bananas are highly dependent on nutrients for their optimal growth and output, accounting for 20-30 per cent of the production costs. Proper nourishment is essential for the production of high-quality banana bunches. Consequently, researchers have studied the nutrient dynamics in banana to improve the efficiency of nutrient utilization. The utilization of biofertilizers is justified due to its positive impact on soil fertility and its ability to support sustainable agricultural production. It is recommended to apply 25 grams each of Azospirillum and phosphorus solubilizing bacteria (PSB) per plant at the time of planting in order to increase production. Additionally, organic sources such as 10 kilograms of farmyard manure (FYM) or 5 kilograms of vermicompost should also be applied at the time of planting. A recommended practice for achieving improved production under a drip irrigation system is to apply 200 grams of nitrogen (N), 40 grams of phosphorus pentoxide (P2O5), and 200 grams of potassium oxide (K_2O) in a split application. Below is the schedule for fertilizer application by soil and drip irrigation developed by Banana Research Station, Jalgaon (Table 10.1).

Table 10.1. Fertilizer Application Schedule through Soil

SN	Time of fertilizer application	N (Urea)	P_2O_5 (SSP)	K_2O (MOP)
1.	Within 30 days after planting	37.5 (82)	40 (250)	50 (83)
2.	75 days after planting	37.5 (82)	-	-
3.	120 days after planting	37.5 (82)	-	-
4.	165 days after planting	37.5 (82)	-	50 (83)
5.	210 days after planting	16.7 (36)	-	-
6.	255 days after planting	16.7 (36)	-	50 (83)
7.	300 days after planting	16.7 (36)	-	50 (83)
Total		**200 (435)**	**40 (250)**	**200 (332)**

Recommended dose of phosphorus is applied through soil at the time of planting.

Fertigation schedule (g/Plant/Week) is depicted in Table 10.2.

Table 10.2. Fertigation Schedule (g/Plant/Week)

SN	Period of application	No. of weeks	N (Urea)	K2O (MOP)
1.	1-16 weeks	16	3.0 (6.52)	2.0 (3.15)
2.	17-28 weeks	12	6.0 (13.0)	5.0 (8.35)
3.	29-40 weeks	12	2.5 (5.45)	4.0 (6.70)
4.	41-44 weeks	4	-	3.0 (5.0)

Recommended dose of phosphorus is applied through soil at the time of planting.

Water Management

As a succulent plant, bananas have a high water need of 1,800-2,500 mm annually. This water can be obtained by well-distributed rainfall or irrigation. Traditionally, farmers have been using the conventional method of flow irrigation to apply irrigation to their crops at intervals of 4 to 7 days, depending on the season and crop growth. In modern times, the majority of banana cultivators are employing the drip irrigation system, a cutting-edge type of irrigation that delivers accurate and controlled amounts of water directly to the root zone. The principle entails the gradual release of water through plastic emitters or drippers using a low-pressure delivery method. The drip irrigation schedule for banana is depicted in Table 10.3.

Table 10.3. The Drip Irrigation Schedule for Banana

Sl. No	Crop growth stage	Duration (weeks)	Quantity of Water (Lit/Plant)
1.	After planting	1-4	4
2.	Juvenile phase	5-9	8-10
3.	Critical growth stage	10-19	12
4.	Flower bud differentiation stage	20-32	16-20
5.	Shooting stage	33-37	20 and above
6.	Bunch development stage	38-50	20 and above

Bunch Management

The male bud is removed after the completion of the female phase. It provides the twin function of preventing the transfer of food into an undesired sink and protecting fruits from thrips assault that may occur underneath the male bud. Removing the 6th or 8th hands resulted in enhanced fruit quality characteristics that met the requirements for export. The Banana Research Station in Jalgaon has suggested applying two sprays of Potassium dihydrogen phosphate (0.5 per cent) + Urea (1 per cent) on banana bunches, 5 and 20 days after the final hand opening. This treatment has been found to enhance the overall quality, maturity, and production of bananas. The clusters are subsequently enclosed with perforated (2 to 6 per cent ventilation), 100 gauge polyethylene bags. These bags serve to shield the clusters from thrips, beetles, dust, rain, cold temperatures in winter, and blistering sun in summer. Additionally, they promote maturation and increase yields.

Intercropping

Intercropping is a widely used technique in banana farms to control weed development, enhance soil fertility, and boost income. During the first phase, it is possible to cultivate several short-duration intercrops in the intervals between plants. During a specific experiment carried out at the Banana Research Station in Jalgaon, it was found that intercropping banana cv. Basrai with groundnut (var. Phule Pragati) during the Kharif season resulted in a good outcome. This intercropping method led to an additional bonus yield of groundnut without negatively impacting the yield of banana. Another trial was conducted where four rows of semi-spreading variety Phule Pandhari of cow pea were sown as an intercrop in Mrig Baugh planting of banana cv. Grand Naine. This resulted in improved outcomes in terms of increased yield and monetary returns. The cultivation of intercrops such as marigold and Sunhemp resulted in a substantial decrease in the nematode population. The Banana Research Station in Jalgaon recommends the use of integrated weed management techniques, such as cross hoeing, weeding, and producing double crops of cowpea, along with incorporating them into the soil, as an effective method for controlling weeds.

Mulching

Mulching has significant implications for the conservation of soil moisture, along with additional benefits such as weed reduction and regulation of the microclimate for plants. Polythene is a highly efficient type of mulch. However, sugarcane trash, paddy straw, dried banana leaves, and sheaths are more cost-effective options. Additionally, they contribute to the soil's organic matter content when used as mulch. An experiment was conducted to evaluate the effects of using wheat straw and banana leaves as mulch material, at a rate of 12.5 kg per plant, in banana orchards. The results showed that the application of mulch at the beginning of the summer season (February) led to an increase in bunch weight and conservation of soil moisture. Similar outcomes have been achieved by applying an 8 per cent kaolin spray on banana leaves as an anti-transpirant during the summer months, starting from February and continuing every 15 days until the beginning of the monsoon season.

Propping

As a result of a dense cluster, plants with bearing may become lodged and uprooted. Thus, the plants are conventionally supported by bamboo, eucalyptus, shevari, or casuarina poles. Placing the supports against the stems on the leaning side of the bunch emergence has the drawback of producing bruising injury to the bunch. Now, polypropylene strips are used to support the pseudostem, replacing the previous method. In this technique, the stripe is securely attached to the neck of the bearing plant, and its two ends are tied to the base of neighboring plants on the side opposite to the direction in which it is leaning. This decreases the expense of supporting and minimizes the occurrence of bruising injuries to the cluster.

Integrated Pest Management

The primary insect pests responsible for causing losses in banana production include the corm weevil, pseudostem borer, red rust thrips, and nematodes. A one-foot-long pseudostem is utilized as the delivery mechanism for the biocontrol agents (Beauveria basiana) and entomopathgenic worms to monitor and manage pests. Polythene sleeves treated with a solution containing 0.1 per cent chlorpyriphos, paraffin oil, and adjuvant effectively reduced thrips infestation by 25-30 per cent. Applying Imidachloprid 0001 per cent by Bell injection (also known as flower bud injection) shortly after the shooting stage has effectively managed the infestation of red rust thrips. Neem-based formulations such as Nimbicidin and Econeem, applied at a rate of 30 ml per plant, or the usage of Pseudomonas flurescence, Pacilmyces lilacines, Verticillium chlamydospermum, and Trichiderma harzianum have been successfully employed for controlling nematodes in banana plants. The application of VAM (specifically Glomus Fasciculatum and Glomus mossaeae) during planting and after a period of 3 months successfully decreased the presence of nematode infestation.

Disease Management

Yellow sigatoka and Cigar end rot are significant fungal diseases, although bacterial illnesses are also noteworthy. Erwinia rot is the primary limitation in the cultivation of Cavendish bananas. Viruses such as Banana Bunchy Top, Infectious chlorosis, and Banana streak continue to pose obstacles to the successful cultivation of bananas. An anonymous study conducted at TNAU, Coimbatore in 2012 created a wireless sensor-based forecasting system for Sigatoka leaf spot disease. The study indicated that relative humidity was the most dependable indicator for predicting the disease. Cultural practices have a substantial impact on the management of Sigatoka sickness. The package comprises a meticulously maintained garden that is free from weeds, clean, and properly drained. Additionally, any diseased plant parts are promptly removed and destroyed, ensuring they are kept away from the garden. As an extra precaution, a single application of fungicide is sprayed in the month of May to prevent any potential fungal infections. Chemical fungicides can be used to control the Sigatoka leaf spot. However, successful management is only achievable when mineral oil (1.0 per cent) is applied, with or without fungicides.

Precision Farming in Mango

Mango is the predominant fruit crop in India, occupying approximately 39.16 per cent of the total land dedicated to fruit cultivation and contributing over 23.09 per cent of the country's total fruit production. Uttar Pradesh has the largest area dedicated to mango cultivation, with 0.24 million hectares. Andhra Pradesh and Bihar follow with 0.20 and 0.15 million hectares, respectively. Andhra Pradesh is the leading producer of mangoes, with an output of 3.07 million tonnes. In comparison, Uttar Pradesh and Bihar produced 2.39 million tonnes and 1.79 million tons, respectively. India has the top position globally in terms of both mango production and the area dedicated to cultivating mangoes. However, its productivity falls far below that of Israel, Mexico, and South Africa. Despite the vast expanse of mango cultivation in India, the per capita availability of mangoes is inadequate. Thus, it is imperative to enhance mango production for direct consumption and processing into diverse goods, catering to both home and international markets.

The production and productivity of mango can be enhanced by effectively and wisely utilizing resources such as water, nutrients, beneficial herbicides, environmentally friendly pesticides, and by expanding cultivation in problematic soils through the development or selection of suitable cultivars and rootstocks. Additionally, implementing high-density plantation and managing the canopy can also contribute to optimization. Efficient post-harvest handling guarantees improved returns by enhancing the shelf-life and quality of the commodity, hence increasing its value.

The current requirement to maintain the commercial viability of orchards is the use of effective ways to enhance mango productivity.

Strategies for Higher Production

Agro-climatic Requirements

Mango exhibits excellent adaptation to tropical and sub-tropical regions. It flourishes in nearly all regions of the country, but it is not economically viable to cultivate it in locations over 600 m. The tree is particularly vulnerable to severe frost, especially during its early stages of growth. Elevated temperature alone does not cause significant harm to mango trees. However, when coupled with low humidity and strong winds, it has a detrimental impact on the tree. Mango cultivars often flourish in regions with an annual rainfall between 75 and 375 centimeters and a distinct dry season. The spatial distribution of rainfall holds greater significance than its total quantity. Arid conditions before to blooming promote abundant flower production. Precipitation occurring during the flowering stage has a negative impact on the crop due to its disruption of the pollination process. While rainfall during the growth of fruit is beneficial, excessive precipitation might harm maturing crops. Severe winds and cyclones during the fruiting season can have a devastating impact by causing a significant amount of fruit to fall prematurely. Ideal conditions for mango cultivation include loamy, alluvial soils that are well drained, aerated, and deep. These soils should also be rich in organic matter and have a pH range of 5.5 to 7.5.

Growing and Potential Belts

Mango is grown in nearly every state of India. The geographical regions where specific crops or industries are thriving in each state are listed below:

State	Growing belts
Andhra Pradesh	Krishna, East and West Godavari, Vishakhapatnam, Srikakulam, Chittoor, Adilabad, Khamman, Vijaynagar
Chhattisgarh	Jabalpur, Raipur, Bastar
Gujarat	Bhavnagar, Surat, Valsad, Junagarh, Mehsana, Khera
Haryana	Karnal, Kurushetra
Jammu & Kashmir	Jammu, Kathwa, Udhampur
Jharkhand	Ranchi, Sindega, Gumla, Hazaribagh, Dumka, Sahibganj, Godda.
Karnataka	Kolar, Bangalore, Tumkur, Kagu
Kerala	Kannur, Palakkad, Trissur, Malappuram
Madhya Pradesh	Rewa, Satna, Durg, Bilaspur, Bastar, Ramnandgaon, Rajgari, Jabalpur, Katni, Balagha
Maharashtra	Ratnagiri, Sindhudurg, Raigarh
Orissa	Sonepur, Bolangir, Gajapati, Koraput, Rayagada, Gunpur, Malkanpuri, Dhenkanal, Ganjam, Puri
Punjab	Gurdaspur, Hoshiarpur, Ropar
Tamil Nadu	Dharmapuri, Vellore, Tiruvallur, Theni, Madurai
Uttaranchal	Almora, Nainital, Dehradun, Bageshwar, Udham Singh Nagar, Haridwar
Uttar Pradesh	Saharanpur, Bulandshahar, Lucknow, Faizabad, Varanasi
West Bengal	Malda, Murshidabad, Nadia

Selection of Suitable Variety

In India, more than a thousand mango varieties are being grown in different parts of the country. A considerable area is under seedling orchards, majority of which is totally neglected, uncared, least productive and without any major economical significance. Most of the Indian commercial cultivars are characteristically specific to geographical adaptation and their performance is satisfactory in a particular region. Therefore, selection of varieties for mango cultivation should be based on their suitability for a particular region.

In north India, varieties can be selected from an array of varieties, well adapted in different regions. In north, Dashehari, Langra, Chausa, Bombay Green and Lucknow Safeda; in south, Banganapalli, Bangalora, Neelum, Mulgoa and Suvernarekha; in west, Alphonso, Kesar, Pairi, Goam Mankurad, Jamadar and Rajapuri; in east, Himsagar, Fazri, Zardalu, Kishanbhog, Gulabkhas and Langra can be selected for commercial plantation.

Newer varieties developed in different parts of the country can be exploited for various regions. Some hybrids, Mallika and Amrapali, have performed well in most of the parts of the country and show wider adaptability. However, Arka Puneet, Arka Neelkiran and Arka Anmol developed from IIHR, Bangalore; Ratna and Sindhu from Vengurla; Ambika from CISH, Lucknow and Neelphanso, Neeleshan Gujarat and Neeleshwari from Paria, Gujarat, can be successfully grown in respective regions. Bangalora, Neelum and Mallika have been recommended for south Telengana region (5). Selection of a cultivars, for specific purpose, is one of the important factors.

Cultivars like Ramkela in Uttar Pradesh and Ashwina in West Bengal are suitable for pickle-making. Alphonso, Dashehari and Mallika are suitable for canning propose.

Variability in clones of commercial varieties like Dashehari, Langra, Neelum, Totapuri and Sunderja have been exploited through making clonal selections. A regular bearing and high-yielding clone, Dashehari 51, has been released by CISH, Lucknow (4). Clonal selections in Langra from Varanasi, Sunderja at Rewa, Neelum and Rumani from Tamil Nadu have been recommended for commercial plantation (10 and 13).

Water-use Efficiency

When creating irrigation schedules, it is important to consider several elements that influence irrigation response, including as soil type, season, region, stage of tree growth, and varieties.

Planting

Planting Material: Mango can be produced either through seed or vegetative propagation methods. Plants are commonly propagated asexually by various processes such as veneer grafting, inarching, and epicotyl grafting.

Planting Season: Planting is often conducted in July-August in rainfed regions and in February-March in irrigated regions. Planting is conducted at the conclusion of the rainy season in areas prone to significant rainfall.

Spacing: The spacing between plants is 10 meters by 10 meters in dry zones and 12 meters by 12 meters in moist zones. The model scheme incorporates a spacing of 8m x 8m, with a population density of 63 plants per acre. This configuration was commonly found in the areas surveyed during the field investigation.

Training of Plants: Pruning young plants during their early growth is crucial for shaping them correctly, especially when the graft has developed branches at a low position.

Nutrition

Fertilizers can be administered in two separate applications, with one half being done immediately after fruit harvesting in June/July, and the other half in October. This practice is suitable for both young and mature orchards, and irrigation should be provided if there is no rainfall. It is advisable to apply a 3 per cent urea solution directly to the leaves of plants grown in sandy soils before to the flowering stage. The table below provides information on the amount of fertilizer used based on the age of the plants.

Age of the plant (in years)	Fertilizer applied
1*	100g. N, 50g. P_2O_5, 100g. K_2O
10	1kg. N, 500g. P_2O_5, 1kg. K_2O
11	-do-

*The doses applied in the subsequent years should be increased every year upto 10 years in the multiple of the first year's dose.

Highly decomposed farm-yard manure can be applied annually. To apply fertilizers in trenches, it is recommended to provide 400g each of nitrogen (N) and potassium oxide (K_2O), and 200g of phosphorus pentoxide (P_2O_5) per plant. Micronutrients can be administered through foliar sprays according to the specific needs.

Irrigation

The frequency and quantity of irrigation required are determined by factors such as soil type, prevailing climate, rainfall patterns, and tree age. Irrigation is unnecessary during the monsoon months, except in cases of prolonged drought.

Age of the plant (in years)/ Growth stage	Irrigation schedule
1	Irrigated at an interval of 2-3 days during dry season.
2-5	Irrigation interval- 4-5 days.
5-8/ fruit set to maturity	Irrigated after every 10-15 days
Full bearing stage	2-3 irrigations after fruit set.

It is not recommended to irrigate frequently in the 2-3 months leading up to the flowering season, since this may encourage the growth of leaves and stems instead of flowers. Watering should be provided when the soil's moisture level reaches 50 per cent of its maximum capacity. Typically, inter-crops are cultivated in the initial stages of plantation, therefore the frequency and mode of watering need to be adapted correspondingly. Basin irrigation is the typical method used for irrigating mango plants. However, the utilization of Drip Irrigation not only diminishes the water demands but also facilitates fertigation in the root zones of the plants.

Intercultural Operations

The frequency and the time of inter-culture operations vary with age of the orchards and existence of inter-crops. The weed problem may not exist immediately after planting the mango crop but it is advisable to break the crust with hand hoe each time after 10-15 irrigations are applied. In case of mono-cropping, the area between the basins should be ploughed at least three times in a year *i.e.* during the pre-monsoon, post-monsoon period and in the last week of November.

Inter-cropping

Intercropping can be used until the mango trees reach an appropriate height and establish a canopy, typically around 5-6 years of age.Intercrops can include leguminous crops such as green gram, black gram, and gram, cereals like wheat, oilseeds like mustard, sesame, and groundnut, vegetable crops such as cabbage, cauliflower, tomato, potato, brinjal, cucumber, pumpkin, bitter gourd, tinda, and lady's finger, as well as spices like chillies. Crops such as pineapple, ginger, and turmeric, which thrive in partially shaded conditions, can be produced in mature orchards. Aside from field crops, it is possible to cultivate some short-duration, less demanding, and compact inter-fillers such as papaya, guava, peach, plum, *etc.*, as long as they do not disrupt the growth of the primary mango crop.It is recommended to cultivate vegetable crops as intercrops in order to achieve higher yields.

Crop Management

Regulation of Bearing: To regulate growth and fruit production, it is advisable to implement appropriate cultural practices such as applying fertilizers and managing diseases and insect pests. One can cultivate common types of mangoes such as Dashehari and Amrapalli. Applying NAA at a concentration of 200 ppm (20 g./100 l. water) during the productive year can promote the regulation of fruit production by stimulating the development of panicles.

Regulation of Fruit Drop: The main factors contributing to fruit drop include embryo abortion, climatic conditions, disrupted water balance, inadequate nutrition, disease and pest infestation, and hormonal abnormalities. Applying a solution of Alar (B-Nine) at a concentration of 100 parts per million (ppm) or 20 ppm, or a solution of 2,4-D at a concentration of 2 grams in 100 liters of water, during either the last week of April or the last week of May, will partially mitigate the occurrence of summer fruit drop in Langra and Dashehari varieties.

Plant Protection Measures

Insect Pests: The commonly observed insect pests include mealy bugs, hoppers, inflorescence midges, fruit flies, and scale insects. To manage these insects, it is advised to use carbaryl, monocrotophos, phosphamidon, and methyl parathion for spraying.

Diseases and Disorders: The crop is susceptible to diseases such as powdery mildew, anthracnose, die back, blight, red rust, and sooty mold. To effectively manage these diseases, it is necessary to carry out the application of suitable pesticides or fungicides, ideally as a preventive measure. If appropriate precautions and control measures are not implemented, disorders can also have an impact on the crop. The main factors contributing to these issues are malformation, biennial bearing, fruit drop, black tip, and clustering. The cultivator should actively seek guidance and expert support in order to avoid and manage infections and abnormalities in the crop.

Pest and Disease Management Schedule

Proposed are monthly schedules for integrated pest and disease management in mango cultivation (6). It is indeterminate.

July

Deep ploughing of orchard after harvesting to expose eggs and pupae of mealy bug and inflorescence midge.

August-September

- ☆ Removal of webs (made by leaf webber) and burning them.
- ☆ If infestation still continues spray carbaryl (0.2 per cent) or monocrotophos (0.04 per cent).
- ☆ Pruning of over-crowded and overlapping branches for control of leaf webber.

October

- ☆ Pruning of infected and dried branches, 10 cm below the dried portion and pasting of copper oxychloride for control of die-back.
- ☆ Spraying of 0.3 per cent copper oxychloride (3g/litre) after pruning for the control of die back, phoma blight, anthracnose and red rust disease.
- ☆ Removal of diseased foliage/twig infected with anthracnose (twig blight phase).
- ☆ Removal of weeds.

November

- ☆ Deep ploughing of the orchards for exposing eggs and pupae of insects and removal of weeds in mango orchard which harbour pests and diseases.
- ☆ Second spraying of copper oxychloride (3g/litre) for control of die-back and foliar diseases.
- ☆ Collection of dropped diseased leaves and burning them.

December

- ☆ Fastening of polythene sheet of 400 gauge thickness, 25 cm wide around the base of tree for controlling mealy bug.
- ☆ Raking of soil around the trunk and mixing with neem cake for management of mealy bug nymphs or apply 2 per cent dust of methyl parathion @ 250 g/tree or 2 per cent chlorpyriphos dust. Application of *Beauvaria bassiana* around tree trunk to manage nymphs of mealy bug.

January

- ☆ Cleaning of polythene bands at regular intervals.
- ☆ Spraying of fenitrothion (0.05 per cent) or dimethoate (0.045 per cent) at budburst stage for control of inflorescence midge.
- ☆ Removal of weeds and infected young leaves of mango for control of powdery mildew.

February

- ☆ First spray of 5 per cent neem seed kernel extract (NSKE) or Nimbicidine (2 per cent) at bud-burst stage for control of hoppers.
- ☆ Spraying of *Verticilium lacani* (106) at bud-burst stage for control of

hopper and it should be repeated during July (second appearance) for controlling nest generation hoppers.

March

- ☆ Second spray of 5 per cent neem seed kernel extract (NSKE) or Nimbicidine (2 per cent) when fruits are at pea-sized stage.
- ☆ First spray of sulphur (2g/litre) to control powdery mildew.

April

- ☆ Third spray of endosulfan (0.07 per cent) if required after 5 days of second spray.
- ☆ Second spray of sulphur (2g/litre) after fruit setting against powdery mildew.
- ☆ Removal of powdery mildew infected leaves and malformed panicles.

May

- ☆ Hanging of fruit fly traps (0.1 per cent methyl euginol + 0.01 per cent malathion) for control of fruit fly June
- ☆ Methyl euginol traps should be continued.
- ☆ Early harvesting of mature fruits to avoid fruit fly infestation.
- ☆ Collection and destruction of fruit fly infested fruits

Harvesting and Post-harvest Management

The optimal time to pick mango fruits is when they reach the mature green stage, ideally in the morning. The maturity stage is determined by evaluating the skin and pulp color, specific gravity, and the number of days elapsed since the fruit was set. The mango cultivars Dashehari and Langra attain maturity 12 weeks after the fruit has been set, whilst Chausa and Mallika require 15 weeks, and Banganapally and Alphonso require 16 weeks. Fruits should be collected either by hand or with a mechanized harvester. It is not advisable to shake branches in order to make them drop, as this can negatively impact their shelf-life. Highly efficient, cost-effective, and easily portable mango harvesters, developed and refined in multiple locations across the country, are already available for use. The harvester enables the collection of fruits while keeping their stalks intact, which enhances their visual attractiveness by preventing the production of unsightly spots on the skin caused by sap burn. By adopting this harvesting method, fruits become more resistant to stem-end rot disease while in storage. Recently, mechanical harvest aids have been developed to tackle the corrosive nature of fruit sap, which can lead to sap burn, as well as the requirement to remove the stalks from the fruits during harvesting and desapping in the packaging process. In order to prevent the abrupt release of sap, fruits are harvested with a fruit stem that is approximately 2-5 cm long. Once the fruits are picked, they are placed in field crates and then transported to a designated packing area where the desapping process occurs. Currently, fruits are harvested by employing hooks to separate them from the panicle, without the presence of a stem. Thus, it is recommended to choose hand harvesting utilizing fruit stalks. The fruit yield of mango is impacted by several factors, such as the mango variety, the tree's bearing characteristics, the climate, the tree's age, the presence of pests and diseases, and the cultural practices employed in the orchard. Grafted trees reach productivity within 3-4 years of being planted. The period of fruit development, commencing at the stage of blooming and concluding at fruit maturity, spans from 100 to 150 days. The age and planting density of a tree are important factors that influence its productivity. However, the quality of fruit is mostly determined by factors such as diversity, cultural norms, and nutrition. The Langra and Chausa cultivars exhibit greater vigor, need a period of 15-20 years to attain their maximum fruit-bearing potential, whilst the Dashehari variety does this in 10 years.

Packaging and Transport

Significant losses of agricultural produce occur throughout the post-harvest handling and distribution process. Timber and bamboo are commonly utilized for manufacturing packaging containers throughout the majority of India. Nevertheless, there are other packaging materials, such as corrugated cardboard cartons, that are readily available in the market and can be utilized as viable substitutes for lumber and bamboo. Tissue paper and polythene foam paper are utilized to encase valuable fresh mangoes.

The post-harvest handling of mangoes encompasses all the processes involved in transporting the fruit from the tree to the table. The significance of these activities lies in their influence on the shelf-life of fruits after harvesting. Factors such as fruit health and quality, timing of harvesting, and post-harvest handling, including considerations such as fruit maturity, color, shape, size, sweetness, position on the tree, presence of pests and diseases, weather conditions, soil moisture, and nutrient availability, all play a role in determining the shelf-life of fruits. These characteristics exhibit variability among individual trees, orchards, and seasons. Therefore, meticulous control of these parameters is essential for achieving improved returns. The exudation of latex following the detachment of the stem flows onto the skin, resulting in a worn-out appearance of the fruits due to the formation of blemishes on the skin. Therefore, the process of fruit harvesting is then accompanied by the steps of washing, removing the sap, cooling, immersing in hot water, applying fungicide, sorting, applying wax, packaging, ripening, transporting, and selling. The level of fruit maturity is a crucial factor throughout the harvesting process, since it significantly impacts the quality and shelf-life of fruits post-harvest.

Chapter 11

Precision Farming of Vegetables

Introduction

Precision agriculture (PA) refers to the utilization of advanced technologies in conjunction with established farming methods to carry out a variety of particular activities on the farm. Site-specific crop management (SSCM), also known as precision agriculture (PA), aims to optimize agricultural operations and inputs to align with the specific differences found in the environment. Unlike the 'all-in' method of applying fertilizers, chemicals, and other inputs, Precision Agriculture (PA) focuses on evaluating the individual requirements of different locations and plants in the field and administering the necessary inputs appropriately.

In addition to improving input management, PA technologies also include several creative methods for harvesting, controlling pests, weeds, and diseases, as well as gaining a deeper understanding of the requirements of vegetable crops, such as irrigation and nutrition. Although PA technologies utilize several innovative technologies to enhance production, decrease costs, and minimize environmental effect, they nevertheless depend on traditional agricultural practices to function. While they may not serve as a universal solution to all difficulties, when employed judiciously, they possess the capacity to effectively handle particular aspects of your expanding enterprise. Commonly employed technology in vegetable production encompass several forms of precision agriculture (PA).

☆ **Global navigation satellite systems or GNSS (commonly known as GPS)** – used as guidance systems for the navigation of tractors, bed formers, and other on-farm machinery.

☆ **Yield mapping/monitoring –** used to understand the variations in crop health in specific areas of a field and provide information for decision-making

☆ **Nutrient/water monitoring** – used to understand the variations in nutrient/water uptake and flow, and provide information for decision-making

☆ **Variable rate controllers** – technology that allows varied amounts of inputs to be applied to specific areas needed, such as water and fertiliser.

Benefits of Using PA in Vegetable Production

Precision agriculture technologies have the capacity to enhance various vegetable production methods by reducing labor costs, improving the accuracy of input application, and providing a deeper understanding of the variations within different areas of your farm. As previously said, it is important to understand that PA is not a universal solution for all problems. Instead, it should be seen as a tool that can help improve the efficiency of your existing systems. Potential advantages of production systems may encompass:

☆ Increased accuracy of bed formation

☆ Reduced compaction

☆ Greater knowledge of drainage patterns

☆ Greater knowledge of soil structure/types

☆ Increased input efficiency (*e.g.* water, fertilisers)

☆ More effective control of pests, weeds, and diseases

☆ Increased consistency of crop development

☆ Increased marketable yield

☆ Increased hygiene standards.

Utilizing precision agriculture (PA) technologies, in conjunction with established agronomic expertise, has the capacity to enhance the productivity and profitability of vegetable enterprises.

Ladyfinger Cultivation Technology

Okra, also referred to as "ladyfinger," is a well favored vegetable among customers. Okra is well regarded among veggies. Okra is mostly composed of protein, carbs, and minerals including calcium and phosphorus, as well as vitamins A, B, C, thiamine, and riboflavin. It provides sufficient quantities of vitamins A and C. Cultivating okra as a cash crop can lead to increased revenues for farmers. Okra fruits have a significant amount of iodine. It is possible to cultivate it in all districts in the region. Agricultural experts have created hybrid okra cultivars to increase productivity and ensure consistent yields of okra, even in different weather conditions. These cultivars demonstrate enhanced resistance to diseases such as yellow vein mosaic virus. Okra cultivation is widespread in warm and tropical places across the globe.

Soil and Field Preparation for Ladyfinger Cultivation

Ladyfinger cultivation flourishes in tropical climates characterized by high temperatures and humidity. An optimum temperature range for seed germination is 27-30 degrees Celsius. This crop is suitable for cultivation throughout both the summer and monsoon seasons. Okra can be grown in several soil types that facilitate effective water drainage. The optimal range for soil pH is typically between 7.0 and 7.8. To achieve consistency, it is recommended to plow the area two to three times, level it, and create ridges.

Optimal Varieties of Ladyfinger

- ☆ Pusa-E4: This is a well developed strain of okra. It demonstrates resilience against aphids and jassids. This plant exhibits tolerance to wilt and yellow vein mosaic virus. The fruits are moderately sized, have a dark green color, and have a length ranging from 12 to 15 cm. The mean crop production is 10 tons per hectare during the summer season and 15 tons per hectare during the monsoon season.

- ☆ Parbhani Kranti: This cultivar exhibits resistance to wilt disease. The process of fruiting typically commences around 50 days after the seeds have been planted. The fruits exhibit a deep green coloration and possess a length ranging from 15 to 18 centimeters. The production varies between 9 and 12 tons per hectare.

- ☆ Arka Anamika: This cultivar was developed by the Indian Institute of Horticultural Research in Bangalore. The plants exhibit a considerable height, ranging from 120 to 150 cm, and possess well-branched stems. The elongated fruit pods provide effortless harvesting. This crop can be cultivated during both seasons and has a productivity of 12-15 tons per hectare.

- ☆ The Seasonal OH-517 cultivar is suitable for planting from December to April. The plant has a moderate height, while the fruits display a dark green color. Due to their lack of spines, they can be easily collected.

Method of Sowing and Seed Quantity

Before planting seeds, it is recommended to plow the field two to three times. To cultivate okra on a yearly basis, it is recommended to have a distance of 40-45 cm between rows and a distance of 25-30 cm between plants within each row. Prior to planting, use a 3-gram dosage of mancozeb carbendazim per kilogram of okra seeds. Partition the entire area into suitably proportioned portions to enhance irrigation. It is recommended to plant okra in raised beds to avoid excessive water accumulation during the rainy season. Under irrigated conditions, a seeding rate of 2.5 to 3 kilogram per hectare is essential, however in unirrigated situations, a higher seeding rate of 5-7 kg per hectare is required. A quantity of 5 kilograms of seeds per hectare is adequate for hybrid varieties. Okra seeds are directly planted in the field.

Irrigation of Ladyfinger

Ladyfinger should be irrigated at intervals of 10-12 days during the month of March, 7-8 days in April, and 4-5 days in May-June.

Preparation of Nursery for Ladyfinger

To prepare the okra nursery, it is recommended to put coco peat in polythene bags using portering. Additionally, the seedling trays should be placed inside a low tunnel. Once the okra seedlings have developed 3-4 leaves, they are ready to be relocated to the field. The optimal depth for nursery portering should range from 2.5 to 3 inches. The temperature must be suitable for the transplantation of okra seedlings. Transplanting okra often occurs during the final week of February.

Pest Management

- ☆ **Aphids and Fruit Borers:** This pest is more prevalent during the rainy season. Its attack on flowers results in premature falling before fruit formation. The fruits become misshapen and unfit for consumption. To control it, spray a solution of 2.5 milliliters per liter of water containing either 25 per cent EC quinalphos, 20 per cent EC chlorpyrifos, or 50 per cent EC profenofos.

- ☆ **Red Spider Mite:** This pest pierces leaf cells with its mouthparts, extracting sap and causing yellowing and distortion of affected leaves. To manage it, mix 2.5 grams of soluble sulphur per litre of water and apply as a spray.

Precision Farming of Brinjal

Brinjal, also known as eggplant (*Solanum melongena*), is a highly popular vegetable crop cultivated in around 500,000 hectares of land in India. It has an average productivity of 16.23 tonnes per hectare. The average yield of brinjal in Rajasthan is 4.38 tonnes per hectare. The state's poor productivity necessitates the implementation of a well defined and exact set of activities that can significantly increase the yield. PFDC Bikaner made attempts to refine and streamline the existing set of practices for brinjal growing. The package includes the incorporation of micro irrigation and fertigation concepts to provide guidance to farmers on irrigation and fertilizer scheduling. The package also includes other components of precision farming, such as nursery growing, selection of appropriate varieties, in-situ moisture conservation, weed control through mulching, plant protection measures, and protection from frost.

Suitable Varieties

- ☆ Long fruits – Pusa Purple Long, Pusa Purple Cluster, Pusa Kranti
- ☆ Round fruits - Pusa Purple Round, H-4, P-8, Pusa Anmol, Pant Rituraj, T-3
- ☆ Hybrid varities - Arka Navneet, Pusa Hybrid -6, F-1 Navkiran

Soil and Climate

Eggplant may be cultivated in several soil types, ranging from sandy loam to heavy clay, regardless of their high pH levels. Light soils are conducive to achieving an early production, whilst clay loam is very suitable for achieving a higher yield. Eggplant necessitates an extended period of warm weather for optimal growth. Fruits undergo distortion due to low temperatures during the chilly season. The most beneficial temperature range for optimum development and yield is a daily mean temperature of 13 to 21°C. The germination of brinjal seeds is optimal at a temperature of 25°C.

Soil Sterilization

Soil sterilization can be accomplished using physical or chemical methods. Solar energy can also be utilized for physical control. Chemical approaches encompass the application of herbicides and fumigants for treatment purposes. UV stabilized 150-micron transparent plastic mulch is suitable for soil sterilization, effectively eliminating hibernating and soil-borne organisms, including plant pathogens and pests. During soil solarization, the transparent plastic sheet allows the incoming solar energy to permeate and be absorbed by the soil.

Raising of Nursery

Prepared are raised beds of 3 meters in length, 1 meter in width, and 0.15 meters in height. Apply 15 kilograms of thoroughly decomposed farmyard manure and 150 grams of single superphosphate. Apply a solution of Captan (3 grams per liter of water) to thoroughly saturate the nursery beds two days prior to sowing, and then smooth the surface. If the seeds have not been treated yet, apply thiram or captan at a rate of 2 grams per kilogram of seed. Plant the seeds at a depth of one centimeter in rows that are spaced five centimeters apart. Encase the seed with a blend of well decomposed manure and fine soil, ensuring it is firmly compacted. Water the plants with a micro sprinkler or a fine rose attachment, both in the morning and evening.

Geometry

The spacing between plants and rows is determined by the characteristics of the soil, the diversity of plants, and the prevailing meteorological conditions. Typically, non-spreading type varieties are spaced at 60x60 cm, whereas spreading type varieties are spaced at 80x60 cm.

An affordable drip irrigation system can be achieved by employing a lateral spacing of 120 cm and dripper spacing of 80 cm for paired row planting at 60 x 80 cm. This system utilizes a 4 lph 16 mm lateral. For sandy soil, it is possible to consider single row planting with a spacing of 80 x 40 cm. This can be achieved by utilizing 16 mm inline drip irrigation with a spacing of 40 cm and a flow rate of 2 liters per hour at a lateral spacing of 80 cm. When choosing the spacing for laterals and drippers, it is important to carefully consider their suitability for the next crop.

Transplanting

The seedlings should be prepared for transplantation within a period of 4 weeks. At this stage, they should have reached a height of approximately 12-15 cm and have developed 4-5 leaves. To strengthen the seedlings, refrain from watering them for a period of 4-6 days before transplantation. Apply a small amount of water to the nursery prior to removing the seedlings. Remove the young plants from the ground without causing harm to their root systems. Transplanting should be conducted in the evening, and it should be followed by irrigation. Exert strong pressure to compact the earth surrounding the young plants.

Irrigation

Apply irrigation to the field according to the specific requirements of the crop. Timely and adequate watering is crucial for optimal growth, blooming, fruit formation, and fruit maturation. The surface irrigation method requires irrigation to be given every third or fourth day during hot weather and every 7 to 12 days during winter. It is important to ensure that the brinjal field is adequately hydrated during periods of frost.

The irrigation is applied on the basis of crop evapotranspiration (ETc=PE x Kp x Kc) considering crop duration (140 days) and Kc value 0.55, 0.8, 1.2, and 0.85 for initial (20 days), crop development (40 days), mid-season (55 days) and late season stage (25 days), respectively.

Drip Irrigation in Brinjal

Drip irrigation is not only water-efficient but also well-suited for fertigation. Significant increases in yield can be achieved by applying optimal amounts of water and fertilizer at the appropriate timing.

Irrigation Schedule for Brinjal Cultivation under Drip Irrigation

Crop Stage	Water applied (lit./ plant)
Establishment (20 days)	1 lit on alternate day
Emergence of new leaf to first flowering	2.8 lit on alternate day
Flowering and fruiting (55 days)	3.00 lit daily
Late season (till last picking)	2.00 lit daily

Manure and Fertilizer

Eggplant is a crop that requires a lot of nutrients, thus it is crucial to apply a well-balanced combination of manure and fertilizer in order to achieve effective crop production. Moreover, the brinjal, being a crop with a lengthy growth period, necessitates a substantial quantity of both manure and fertilizer. It is recommended to incorporate 15 tonnes of well-rotted FYM during field preparation. The crop requires an additional 80 kg of nitrogen (N), 80 kg of phosphorous (P_2O_5), and 60 kg of potassium (K_2O). Apply half the recommended amount of nitrogen (N) and the full recommended amounts of phosphorus (P) and potassium (K) as a basal dose. Apply the remaining half dose in two equal splits: one at 20 days after sowing (DAS) and the other at the time of blooming. In order to practice fertigation, it is recommended to apply the fertilizer in ten separate doses, with an interval of 10 days between each application. This should be done using appropriate water-soluble fertilizers such as 19-19-19, 12-61-00, and urea.

Inter-Culture and Weed Control

Weeds must be promptly managed during the vigorous growth of crops by chemical or mechanical means. Hand weeding is the optimal and cost-efficient approach for managing weed growth. Incorporating fluchloralin into the soil before planting (at a rate of 1-1.5 kg/ha) and spraying alachlor on the soil surface before planting (also at a rate of 1-1.5 kg/ha) has been proven to be an effective method for weed control. The application of 25-micron black plastic mulch in brinjal cultivation suppresses the germination and growth of weeds while also preserving soil moisture. Additionally, it aids in regulating soil temperature.

Plant Protection

(A) Insects

The shoot and fruit borer is a kind of insect known as Leucinoides orbanalis. The tiny larva infiltrates the plant tissues. During the first phase, it mostly targets the terminal shoots and subsequently affects the immature fruits. The young shoots and leaves exhibit signs of withering and shedding as a result of insect infestation.

Management

- ☆ Rouge out the affected plants and destroy them.
- ☆ Spray carbaryl (0.1 per cent) or endosulfan (0.05 per cent) as soon as attack is seen and repeat the spray after 15 days
- ☆ Apply carbofuran 3G or phorate 10 G @ 1.5 kg a.i./ha in the soil before sowing and transplanting.

Jassids. They extract the cell sap from the underside of the leaves. Impacted plants exhibit a pale yellow hue. Jassids are also vectors of the virus that causes small leaf disease in brinjal.

Management

Spray carbaryl (0.1 per cent) or endosulfan (0.05 per cent) or malathion (0.1 per cent) as soon as attack is seen and repeat the spray after 10 days.

(B) Diseases

Damping off. (Pythium spp., Phytophthora spp., Rhizoctonia spp.). The fungus infects seedlings at both the pre and post-emergence stages. The impacted seedlings experience a collapse in the collar region, ultimately leading to their demise.

Management

- ☆ Avoid over watering.
- ☆ Drench the nursery bed with thiram or captan @0.4 per cent at 5-7 days after germination.
- ☆ Drench the soil with formalin (7 per cent) upto 15 cm depth.
- ☆ Treat the seed with captan or thiram @ 3 g/kg of seed.

Viral Disease

Mosaic. The leaves of the affected plants have mottling, characterized by elevated dark green patches. Blister formation occurs on the leaves, resulting in a reduction in leaf size. Aphids serve as vectors for the transmission of the virus.

Management

- ☆ Collect the seeds from virus free plants.
- ☆ Rogue out the infected plants from the field.
- ☆ Spray dimethoate (0.05 per cent) or malathion (0.05 per cent) at 10 days interval.

Harvesting and Yield

Harvesting should be conducted when the fruits have reached their maximum size and color, but before the onset of ripening. The optimal stage for picking fruits is characterized by tenderness, vibrant color, and a glossy appearance. By implementing the precision farming package of methods, it is possible to obtain a yield level of 300 qt/ha or more.

Chapter 12

Precision Farming in Ornamental Crops

Introduction

Precision farming, once primarily associated with staple food crops, is increasingly finding application in the cultivation of ornamental plants. This innovative approach leverages advanced technologies such as sensors, drones, and data analytics to optimize every facet of cultivation, from soil management and irrigation to pest control and nutrient supplementation. Ornamental crops, which encompass a vast array of flowers, shrubs, and decorative plants, present unique challenges and opportunities in precision farming. By precisely tailoring inputs and interventions based on real-time data and analytics, growers can enhance yield quality, reduce environmental impact, and ensure sustainable production practices. As the demand for high-quality ornamental crops grows alongside environmental concerns, the integration of precision farming techniques promises to revolutionize this sector, ensuring both economic viability and ecological sustainability in ornamental crop cultivation.

Precision Production Technology of Jasmine

Jasmine is a significant commodity in the loose flower industry. TNAU, Coimbatore, has created a cutting-edge precision production technique to enhance the production, productivity, and net revenue of producers. The technology encompasses meticulous field preparation, accurate spacing, the use of enriched medium consortia, the delivery of fertilizers using drip and fertigation systems, the spraying of biostimulants and micronutrients, and the implementation of integrated pest and disease management practices. The implementation of the technique leads to a 65 per cent augmentation in bloom production.

Jasmine is the predominant and economically significant traditional flower crop in southern India. Since ancient times, it has been regarded as a sacred flower. The plants can be cultivated as either shrubs or climbers. Flower buds serve the purpose of creating garlands, wreaths, adorning women's hair, and being used as religious offerings. Precision farming has evolved as a solution to mitigate the negative consequences of excessive or insufficient use of inputs. Site-specific management techniques have been devised to optimize crop output and limit environmental pollution and degradation, hence promoting sustainable development. Precision farming is a method that involves accurately measuring and applying the right amount of nutrients and water to crops at specific stages of growth. This is done to ensure that the crops receive adequate nourishment and maximize yield, while minimizing any negative impact on the land and environment.

Technology

Field Preparation

To provide optimal soil texture in the root zone and efficient drainage during the wet season, a tractor-drawn chisel plow is used to plough the field. Afterwards, three ploughings are carried out using a disc, cultivator, and two rounds with a rotavator. Following the process of leveling, it is necessary to build a drip system in the designated area. Subsequently, water is propelled by a 10 horsepower engine and directed to the main pipeline following filtration via a screen filter. The water is transported from the source to the field using a main line consisting of 2.5" PVC pipes. A fertigation tank with a capacity of 60 liters has been installed for the purpose of fertigation. 2.0" PVC pipes are installed as sub-main pipes, branching off from the main pipes. From the sub-main, laterals with a diameter of either 12 mm or 16 mm are installed to run alongside each row of plants. Emitters are positioned at 50 cm intervals along the laterals, with a discharge rate of 4.0 litres per hour. The planting is conducted with a row spacing of 1.2 meters and a plant spacing of 1 meter.

Spacing

A planting arrangement with a spacing of 1.2m × 1.0m is utilized, allowing for a total of 8,300 plants to be accommodated inside one hectare. The adoption of closure spacing results in an increase in plant population per unit area, hence enhancing crop productivity.

Basal Application

In addition to applying 5kg of FYM per pit, 500g of neem cake per pit, and 100g of vermicompost per pit, 3kg of Azospirillum and 3kg of Phosphobacteria per hectare are also applied. The utilization of enhanced growth media, such as vermicompost and neem cake, promotes improved crop establishment.

Fertilizer Application

The recommended dose of fertilizers is applied entirely through fertigation, using water-soluble fertilizers such as polyfed (19:19:19), KNO3 (13:0:45), urea, mono ammonium phosphate (12:61:0), and sulphate of potash. This application is done in splits at weekly intervals. Approximately 75 per cent of the appropriate amount of phosphorus (P) is provided as single superphosphate (SSP) by basal soil treatment.

Fertigation is the most efficient technique of applying fertilizer, as it reduces losses due to leaching and volatilization. This method improves the efficiency of nutrient consumption and minimizes contamination of groundwater. Additionally, fertigation allows for precise adjustment of water and nutrient supplies to match the specific needs of the crop. Additionally, it enables the precise delivery of nutrients straight to the root zone at crucial times when there is a high demand for nutrients.

Irrigation

The customary technique of cultivation involves the use of surface or flood irrigation, which is applied on a weekly basis. The precision method of agriculture involves adopting drip irrigation once every three days. Drip irrigation maximizes water efficiency by effectively utilizing every drop of water. Additionally, it promotes optimal crop growth and higher yields by consistently maintaining a stable moisture level in the crop's root zone through frequent irrigation at shorter intervals.

Cymes per plant in precision method and Cymes per plant in conventional method are presented in Figures 12.1 and 12.2 respectively.

Figure 12.1. Cymes per Plant in Precision Method.

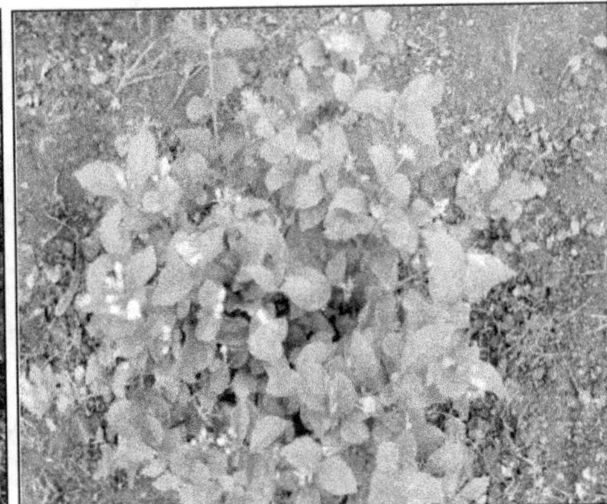

Figure 12.2. Cymes per Plant in Conventional Method.

Biofertilizer

In conventional systems, biofertilizers are not utilized. Each biofertilizer, Azospirillum and phosphobacteria, is sprayed at a rate of 3 kg per hectare. Both stimulate or improve the growth, development, and production of crops by fixing nitrogen, solubilizing phosphorus, and synthesizing growth-promoting substances such as IAA, GA, and cytokinins.

Biostimulant Spray

To get high-quality flowers, improved yield, and flowers during the off-season (December - February), it is recommended to use a spray containing Panchagavya at a concentration of 3 per cent and humic acid at a concentration of 0.4 per cent on a monthly basis. Panchagavya and humic acid are organic biostimulants derived from living creatures. They are rich in beneficial microbes, growth hormones, and essential nutrients that promote plant development and increase crop production.

Buds per cyme in precision method and Buds per cyme in conventional method are presented in Figures 12.3 and 12.4 respectively.

Figure 12.3. Buds per Cyme in Precision Method.   **Figure 12.4. Buds per Cyme in Conventional Method.**

Micronutrient Spray

In addition to a solution containing 5 grams of Fe_2SO_4 per liter, a solution containing 5 grams of $ZnSO_4$ per liter is also applied in order to achieve high-quality flowers. Iron and zinc are essential micronutrients that are crucial for chlorophyll synthesis. The administration of these micronutrients reduces chlorosis symptoms in leaves, thus enhancing the quality of flowers.

Pest Management

In the typical cultivation system, bud worm is controlled by spraying Monocrotophos at a rate of 2ml per liter. However, in the precision system, we apply Thiochloprid 240 SC at a concentration of 0.5 per cent as an insecticide that is quickly biodegradable and has low residual toxicity. Additionally, for controlling gall midge, we use Rynaxypyr 20 SC at a concentration of 0.5ml per litre.

Disease Management

Carbandazim at a concentration of 0.2 per cent or Dithane M45 at a concentration of 0.1 per cent is employed to manage leaf spot in the traditional method of farming. However, in the precision system, the management of Cercospora leaf spot involves applying Bacillus subtilis to the soil at a rate of 25g/m2 during planting, and then applying B. subtilis to the leaves at a concentration of 0.5 per cent every month. To control Alternaria leaf blight, Pseudomonas fluorescens should be applied to the soil at a rate of 25 g/m2 after planting, and then applied to the leaves at a concentration of 0.5 per cent every month. The recently standardized environmentally-friendly techniques for managing bud worm and gall midge can reduce both the amount and frequency of pesticide and fungicide applications, while also minimizing residual toxicity.

Yield

The conventional cultivation approach typically yields an average of 8 tonnes per hectare, while the concrete recovery rate ranges from 0.18 per cent to 0.19 per cent. However, by using a precise system, the yield can be increased to 12 - 13 tons per hectare, while achieving a concrete recovery rate of 0.21-0.22 per cent.

Bougainvillea

Bougainvillea is a genus of spiny decorative climbers, shrubs, or trees. The inflorescence is composed of prominent sepal like bracts that encircle three uncomplicated waxy blooms. This species is indigenous to South America, specifically ranging from Brazil in the east to Peru in the west, and extending southward to southern Argentina.

The vine species have a height range of 1 to 12 meters (3 to 40 feet) and climb over other plants using their sharp thorns. They remain green throughout the year in regions with continuous rainfall, but shed their leaves during dry seasons. The leaves exhibit an alternate arrangement and have a basic ovate-acuminate shape, measuring 4–13 cm in length and 2–6 cm in width. The plant's actual flower is diminutive and typically white, but each group of three blossoms is encircled by three or six bracts exhibiting the vibrant hues commonly associated with the plant, such as pink, magenta, purple, red, orange, white, or yellow. Bougainvillea glabra is commonly known as "paper flower" due to its thin and papery bracts. The fruit is a slender achene with five lobes. Bougainvillea flowers having showing bracts, 2 Bougainvillea leaves is given Figure 12.5.

Cultivation and Uses

Bougainvillea are widely cultivated as decorative plants in regions characterized by warm weather, like Florida, South Carolina, and the Mediterranean Basin. While bougainvillea is susceptible to frost and thrives in USDA Hardiness Zones 9b and 10, it can also be utilized as a houseplant or placed in a hanging basket in colder regions. It is a highly suitable plant for hot seasons in the landscape, and its ability

Figure 12.5. Bougainvillea Flowers having Showing Bracts, 2 Bougainvillea Leaves.

to withstand drought makes it well-suited for warm areas throughout the year. Due to its strong salt resistance, it is an ideal option for providing color in coastal settings. It can be trimmed into a traditional form, but it is also cultivated along fence lines, on walls, in containers and hanging baskets, and used as a hedge or an accent plant. The plant has lengthy, curving branches covered in thorns. It produces heart-shaped leaves and abundant papery bracts in various colors such as white, pink, orange, purple, and burgundy. Numerous varieties, including those with double-flowered and variegated characteristics, are currently accessible. A significant number of the bougainvillea plants found today are the outcome of crossbreeding between just three out of the eighteen South American species acknowledged by botanists. Presently, the global population of bougainvillea has more than 300 distinct kinds. Due to multiple generations of crossbreeding, it is challenging to determine the specific origins of many hybrids. Spontaneous natural mutations are observed worldwide, particularly in areas where there is a high production of plants, leading to the occurrence of budsports. As a result, there have been numerous designations for the identical cultivar (or variety), contributing to the perplexity around the nomenclature of bougainvillea cultivars. The growth rate of bougainvillea ranges from sluggish to quick, contingent upon the specific type. In equatorial locations, they exhibit a tendency to bloom continuously throughout the year. In other locations, the bloom cycles of these flowers usually last for four to six weeks and occur throughout specific seasons. Bougainvillea thrive in arid soil, in intense sunlight, and with regular fertilization. However, after they are established, plants require minimal watering and may not thrive if excessively irrigated. Tip cuttings provide a simple and efficient method for propagating them. Bougainvillea is highly appealing to Bonsai lovers because of its effortless trainability and its bright springtime blossoms. Indoor houseplants can be maintained in temperate regions and controlled in size using bonsai techniques.

B. × buttiana is a garden hybrid of B. glabra and B. peruviana. It has produced numerous gardenworthy cultivars.

Bougainvillea are relatively pest-free plants, but they may be susceptible to worms, snails and aphids. The larvae of some Lepidoptera species also use them as food plants, for example the giant leopard moth (Hypercompe scribonia).

Culture

Bougainvillea (Bougainvillea Commers.) is a widely cultivated ornamental shrub known for its vibrant and appealing bracts. It belongs to the Nyctaginaceae family. It originates from the tropical region of South America. The introduction of this item to India occurred in 1860, originating from Europe. The great adaptability of this product to various agroclimatic situations. The commercial significance of bougainvillea in the nursery trade is due to its ability to blend with a wide range of repeating flowering habits. There are approximately 16 species within the plant. Among the total of 16 species, 4 possess significant ornamental value. There are over 1000 cultivars belonging to several species that are available worldwide. Among them, only three species, namely B. spectabilis, Wild B GlabraChoicy, and B.peruviana Humb and Bompossess, exhibit vibrant coloration for their decorative appeal.

Watering

Bougainvillea exhibits a high tolerance for drought, and it is advisable to adapt irrigation practices to maintain a somewhat dry environment. These plants are susceptible to excessive irrigation but should not be left entirely dry.

Fertilizer

To achieve optimal outcomes, it is recommended to utilize organic fertilizer additives or controlled-release fertilizers in order to regulate the release of nitrogen. When planting, enhance the soil by incorporating a fertilizer that has a high concentration of phosphate. In order to maintain a culture for a long period of time, it is necessary to apply a controlled-release fertilizer as a topdressing. Avoid excessive fertilization. Excessive use of fertilizer will stimulate the growth of leaves and stems while preventing the formation of flowers. Bougainvillea requires regular fertilization using formulas with NPK levels of either 1:1:1 or 2:1:2. The utilization of soluble trace elements aids in the prevention of leaf chlorosis. Micronutrient treatments can be administered at half the recommended dosage, twice annually.

Pruning

Bougainvillea exhibits a favorable reaction to pruning. If bougainvillea is not trimmed periodically, it can become a dense and tangled mixture of old and new growth, which can result in overcrowding and make it more susceptible to pests and diseases. To mitigate overpopulation, remove any superfluous shoots. Prune all side branches, trimming them to a length of two or three buds from the main stems. These will produce the new flowers and bracts. Pruning is essential for shaping and guiding the growth of plants due to the aggressive growth of shoots. Flowers are produced on newly formed plant parts, so it is required to trim and prune in order to stimulate the formation of new plant parts. Pruning should be conducted once flowering has concluded, as this stimulates the emergence of new growth, which will give rise to the subsequent wave of flowers. To decrease the dimensions of plants, trim them by approximately one-third, eliminating any thin and twiggy growth. Remove the unwanted shoots that grow from the base of the plant in order to promote growth at the top. Decaying timber should be promptly eliminated upon its emergence. The lengthy branches can be cultivated into different forms and elevations, such as espaliers, arbors, contorted or interwoven trunks, or even sizable, imaginative creatures.

Controlling Timing of Flowering

Environment

Bougainvillea will bloom earlier and more abundantly when exposed to high levels of light, moderate temperatures, and extended periods of darkness. Short day-lengths promote or stimulate the process of blossoming. The day-lengths range from 8 to 11 hours, with a strong light intensity and temperatures above 58 to 64°F. Excessive shadow hampers the process of blossoming. Under prolonged periods of drought, plants may experience a phenomenon known as drought stress, which can trigger the process of flowering, even when exposed to extended periods of daylight. Growers often intentionally allow plants to reach the point of wilting in order to stimulate flowering. Exercise caution, since an excessive amount of drying might lead to the shedding of leaves and the induction of dormancy.

Cultural Practices

Excessive pruning of plants will hinder the process of flowering. Excessive application of fertilizer will stimulate plant development but impede the process of flowering. Nitrogen and phosphate are essential for the process of flowering, although it is important to avoid excessive fertilization. Maintain soil moisture at a low level.

Pot Culture

Light

Plants should be positioned in an area with ample illumination or in close proximity to a window that receives a minimum of 4000 foot-candles of light. Due to their high light requirement, these plants tend to shed their leaves in low-light interior conditions. Bracts cultivated in partial shade will exhibit a paler hue compared to those produced under direct sunlight. For comparison, the intensity of light at midday on a cloudless summer day is roughly 10,000 footcandles. During a cloudy winter day, the level of illumination may drop to as low as 500 foot-candles. When inside a building, during a sunny summer day, the amount of sunlight that comes in via a window can range from 4000 to 8000 foot-candles. However, the level of sunlight in the shaded areas next to the window will only be around 600 foot-candles. Indoor natural light on the shaded side of a home can range from 150 to 250 foot-candles, which is influenced by factors such as window area and the presence of eaves, window blinds, or curtains.

Media

Any potting media with good drainage is appropriate for cultivating bougainvillea. A peat:perlite medium with a 1:1 ratio by volume is appropriate. If additional weight is required to balance the pot, artificial sand or soil can be added. For optimal root development and flowering, it is crucial to ensure that the medium has good drainage. Refrain from using medium that have a high concentration of peat and possess strong water retention capabilities. These media varieties have a high water retention capacity, which can lead to the development of root rot. The optimal pH range for the media should be between 5.5 and 6.0.

Watering

Plants should be irrigated when the top layer of the growing media dries out. Close vigilance is required as plants have a propensity to deplete the accessible moisture in the containers. The water requirements vary based on the medium type, ambient circumstances, plant size, and pot size. Providing ample water, but at longer intervals, is more beneficial than frequent, shallow waterings.

Fertilizing

For pot growing, it is recommended to use a controlled-release fertilizer with a balanced nutrient ratio (*e.g.*, 8-8-8 or 10-10-10) and apply it every three months. Water-soluble fertilizer formulations can be applied either once a week or once every two weeks, using a diluted solution that contains only half of the normal nutrient concentration.

Pruning

Regular pruning is essential for preserving the proper dimensions and form of the plant, while also minimizing the development of new growth that is tender and thorny. Regular pruning stimulates continuous regrowth and blooming. Trim juvenile plants to promote the development of a sturdy structure of vigorous shoots originating from the plant's lower portion. Trim and sculpt plants once they have bloomed, maintaining a height of approximately 3 feet and eliminating any thin and wiry growth. Remove any feeble or impaired vegetation.

Growth Regulators

To promote lateral branching, the plant tips can be pulled back and applications of BA (benzyladenine) at a concentration of 50–100 ppm can be made. Administer a single spray 24 hours following the initial pinch, and another spray 24 hours after the second pinch. Using 2 ounces of dikegulac sodium per gallon can replace the second pinch-plus-BA application to enhance branching. Atrimmec® is a commercial product that contains dikegulac sodium as its active component, with a concentration of 18.5 per cent. It is typically used at a rate of 1 ounce per gallon. Apply sprays to unpinched shoots when they reach a length of 3 inches or to trimmed plants three days after pruning. Refrain from administering treatment to plants that are experiencing stress.

In order to inhibit growth, Cycocel® (chlormequat) has been applied to potted bougainvillea as a soil drench at a rate of 0.01-0.02 ounce per pot, specifically when the axillary buds begin to enlarge after the initial pruning. A-Rest™ (ancymidol) and Bonzi® (paclobutrazol) are both highly efficacious. Apply paclobutrazol at a concentration of 20-40 parts per million (ppm) when the plants have reached a size suitable for sale.

To inhibit the shedding of bracts, apply a solution of NAA (naphthaleneacetic acid) in the form of a spray or dip, with a concentration of 10-30 parts per million. At a higher rate, NAA can stimulate the shedding of undeveloped bracts.

Prior to widespread use, it is advisable to conduct preliminary tests of plant growth regulators on a limited number of plants. It is important to consistently adhere to the guidelines provided on the label. If the rates or techniques for applying labels are different from those provided above, adhere to the instructions on the label.

Propagation

Cuttings

Propagation can be achieved using softwood terminals, green wood in the process of maturing, and stem sections of intermediate wood that have reached maturity. The stem cuttings should have a minimum thickness of 1/8 inch and should possess a minimum of three to five nodes. During the rooting process, it is OK to keep the leaves on the cuttings. However, it is important to remove any leaves that are submerged in the rooting medium. Utilize a thoroughly drained rooting media, such as a mixture of peat and perlite in equal volumes. Alternative substrates such as artificial sand and peat or coir (coconut fiber) are effective for roots. Place the cuttings into the medium, inserting them to a depth of 1-2 inches, and ensure that they are watered properly. Rooting cuttings can be done by directly placing them in small pots or Jiffy-7s. Multiple cuttings can be propagated together in larger pots with a diameter of 5–6 inches. Additionally, foam propagation blocks can be utilized. Rooting hormone is not necessary for softwood terminals of cultivars that are easy to root. When dealing with older wood, it is typical to employ a rooting hormone called IBA (3-indolebutyric acid) at a concentration of 2000-6000 ppm. Greater concentrations may be required for cultivars that are more challenging to propagate through rooting. Intermittent misting is a frequently employed technique to avoid dehydration while promoting root growth. It is imperative to prevent cuttings from wilting. The duration required for rooting ranges from 4 to 12 weeks, depending on the specific cultivar and the application of a rooting hormone. Exercise cautious when transplanting young plants to prevent any harm to their delicate roots. Applying a fungicide that targets a wide range of fungi as a soil treatment during the first planting of cuttings and again after transplanting can effectively prevent root rot.

Leaf-bud Cuttings

Leaf-bud cuttings are a suitable option when there is a scarcity of source material. Each node can serve as a cutting. The cutting, derived from partially grown shoots, comprises a leaf blade and a short segment of the stem (1–11 D2 inches) with the accompanying axillary bud. Trim the stem part approximately 1/2 to 1 inch above and below the point where the leaf is attached. Position the bud in a vertical orientation within a rooting medium and gently conceal it with a layer of 1/4 inch, so that just the leaf blade is visible.

Grafting

Certain cultivars lacking chlorophyll in their leaves provide challenges when attempting to reproduce them from cuttings. In order to successfully propagate these cultivars, they must be grafted onto a robust rootstock. Grafting is beneficial for delicate cultivars that possess vulnerable root systems. It is also employed when there is a need to have various cultivars on a single plant. The scion must be devoid of any ailment. The rootstock may consist of either a seedling or a rooted cutting obtained from a pre-existing and well-established plant. After making the join, it is important to coat all sliced surfaces with grafting wax in order to avoid any loss of moisture. Eliminate the sprouts from the rootstock. Place the grafted plant in a moist atmosphere to prevent desiccation of the scion. Several types of grafts can be utilized, such as wedge, whip or tongue, or approach graft.

Seeds

Seeds rapidly germinate and do not require any treatments to end dormancy.

Pests and Diseases

Aphids

Aphids are little, pliable insects equipped with elongated, flexible mouthparts which they employ to puncture stems, leaves, and other delicate plant structures in order to extract plant juices. Bougainvillea, like most plants, is susceptible to infestation by one or more kinds of aphids, which may feed on it from time to time.

Aphids exhibit a range of colors, including green, yellow, brown, red, or black, which is contingent upon the specific plants they consume. The green aphid is typically observed feeding on the delicate fresh tissue of bougainvillea. These insects are all tiny and have a pear-shaped body, long legs, and antennae. Minimal to moderate populations of aphids that feed on leaves typically do not cause harm in gardens or on trees. Nevertheless, the presence of substantial populations leads to the occurrence of curling, yellowing, and deformation of leaves, as well as the stunting of shoots. In addition, they have the ability to generate substantial amounts of a viscous secretion called honeydew, which frequently becomes darkened due to the proliferation of a fungus known as sooty mold.

☆ **Biological control:** or natural enemies can be very important in the control of aphids.

☆ **Cultural Control:** before planting bougainvillea, check surrounding areas for sources of aphids and remove them.

☆ **Chemical Control:** Insecticidal soap, neem oil, and narrow-range oil (*e.g.,* supreme or superior parafinic-type oil) provide temporary control if applied to thoroughly cover infested foliage.

Leaf Miner

Leaf miners feed by constructing superficial tunnels, known as mines, in the tender foliage of bougainvillea shrubs. The life cycle of leaf miners consists of four distinct stages: egg, larva, pupa, and adult moth. Adults do not harm plants and have a lifespan of only 1 to 2 weeks. The eggs typically undergo hatching around one week after being deposited by the adult moth. Upon hatching, the just emerging larvae promptly commence feeding within the leaf and initially generate minuscule, practically imperceptible tunnels. As the larva matures, its sinuous trail of mines becomes increasingly conspicuous. The bougainvillea bushes are covered in numerous intricate pathways like mines, making it nearly impossible to locate the adult insects.

Leaf miners can only thrive as larvae in the delicate, youthful, glossy leaf flush of bougainvillea. Mature leaves that have become rigid are resistant unless there is an exceptionally large number of pests. The larvae burrow into the lower or upper surface of recently sprouted leaves, resulting in their deformation and curling. The leaves do not regain their original shape and appearance, resulting in an unattractive appearance when the plant is heavily infested.

Place pheromone traps at a height level that is approximately the same as the shoulders on the plant that is being affected. Adhere to the manufacturer's guidelines for maintaining the trap, including the recommended frequency for replacing the pheromone dispenser.

Bacterial and Fungal Leaf Spot

These typically occur when environmental conditions are conducive, such as when there is an excessive accumulation of moisture on the leaves of bougainvillea for a prolonged period. Lesions will form either at the outer edge of the leaf or in the spaces between veins and will then expand outward. Over time, the edges of the leaves may become uneven and frayed as the dead tissue dries out and becomes thin like paper. Puckered and deformed growth occurs when developing leaves and bracts become infected. Prevention of this particular infection can be accomplished by refraining from contact with damp leaves of the plants. Trim the big branches by cutting them back and creating distance between them, especially if they are overlapping. Eliminate diseased leaves and plants from the cultivation area and ensure the complete destruction of the remains.

If the illness becomes more challenging, one may need to use chemical sprays. It is crucial to strictly adhere to the instructions on the package.

Scale Insects

Armored scales, belonging to the family Diaspididae, has a compressed, plate-shaped protective covering that is less than 1/8 inch in diameter. The physical body of the insect is located beneath the protective covering. The covers frequently include a distinctively colored, subtle protrusion. Concentric rings are created as each nymphal stage (instar) produces a growth to its protective covering. Armored scales do not produce honeydew. Destructive species encompass the Florida red scale and Oriental scale. Soft scales might have a smooth, cottony, or waxy texture and are no longer than 1D4 inch. Armored scales are often smaller and have a flatter and less rounded shape compared to them. Their surface constitutes the integument of the insect and is non-removable. Soft scales extract nutrients from the phloem tissue of plants and release a copious amount of honeydew, a sweet liquid that drips from their bodies. Soft scales encompass black scale, brown soft scale, and mealie bug. Many scale species exhibit a form of reproduction known as parthenogenesis, in which females are capable of reproducing without the need for mating due to the absence of males. Upon reaching maturity, adult females generate eggs that are often concealed beneath their body or protective covering. Eggs undergo hatching and transform into small, mobile first-instar nymphs, typically exhibiting a yellow to orangish coloration in the majority of species. Crawlers traverse the plant surface, are dispersed by wind to adjacent plants, or can be unintentionally transported by humans or avian creatures. They establish themselves and commence feeding within one to two days after emerging. Severe infestation of scales on plants can cause the leaves to appear wilted, turn yellow, and fall off prematurely. Scales may induce leaf curling or result in malformed blooms and fruit. Parasitic wasps and predators such as beetles, bugs, lacewings, and mites frequently regulate the population of scales. Horticultural oils are petroleum products that have been refined to a high degree, and are commonly referred to as narrow-range, superior, or supreme oils. Additionally, there are botanical oils that can be obtained from plants. Applying oil alone to plants at the appropriate time is typically sufficient for effective control.

Thrips

Thrips are little, elongated insects with wings that have fringed edges. They obtain nourishment by piercing their host and extracting the cellular contents. Some species of thrips are advantageous predators that exclusively consume mites and other insects. Pest species refer to organisms that consume plants and cause damage by scarring the surfaces of leaves, flowers, or fruits, or by distorting plant components.

The majority of mature thrips are slim, tiny (less than 1/20 inch in length), and possess elongated fringes on the edges of both pairs of their elongated, narrow wings. Immature thrips, also known as larvae or nymphs, have a similar body form with a long and slender abdomen, but they do not have wings. Thrips exhibit a wide spectrum of colors, varying from transparent white or yellowish to dark brown or blackish, which is contingent upon their species and stage of life.

Thrips have a preference for feeding on tissue that is experiencing rapid growth. Thrips feeding usually results in little scars on leaves and fruit, known as stippling, and might hinder growth. Western flower thrips mainly infest herbaceous plants, although they can also cause harm to woody plants like roses by damaging its constantly or late-blossoming flowers, particularly when numbers are large. While thrips damage to leaves may be visually unappealing, their activity typically does not justify the application of pesticide sprays. Systemic insecticides can be used by a certified pesticide applicator for ornamental nonfood plants, whereas oils are effective in controlling adult pests.

Spider Mite

Mites are common pests in landscapes and gardens and can be found feeding on many fruit trees, vines, berries, vegetables, and ornamental plants. Although related to insects, mites are not insects but members of the arachnid class along with spiders and ticks.

To the naked eye, spider mites look like tiny moving dots Spider mites live in colonies, mostly on the under-surfaces of leaves; a single colony may contain hundreds of individuals.

Mites cause damage by sucking cell contents from leaves. A small number of mites is not usually reason for concern, but very high population levels high enough to show visible damage to leaves can be damaging to plants, especially herbaceous ones. At first, the damage shows up as a stippling of light dots on the leaves; sometimes the leaves take on a bronze color. As feeding continues, the leaves turn yellow and drop off. Often leaves, twigs, and fruit are covered with large amounts of webbing. Damage is usually worse when compounded by water stress.

Spider mites have many natural enemies, which limit their numbers in many landscapes and gardens, especially when undisturbed by pesticide sprays. Cultural practices can have a significant impact on spider mites. Dusty conditions often lead to mite outbreaks. Apply water to pathways and other dusty areas at regular intervals. Spider mites frequently become a problem after the application of insecticides. Such outbreaks are commonly a result of the insecticide killing off the natural enemies of the mites.

Slugs and Snails

Snails and slugs propel themselves by smoothly sliding on a muscular appendage known as a "foot." The muscle continuously releases mucus, which subsequently dries to create the shiny "slime trail" that indicates the existence of either insect. Both slugs and snails are hermaphrodites, meaning they possess both male and female reproductive organs, allowing them all to have the capability to lay eggs. Adult brown garden snails deposit approximately 80 spherical, pearly white eggs at once into a cavity in the top layer of soil. They have the potential to lay eggs up to six times year. The maturation process of snails typically spans approximately 2 years. Slugs achieve maturity within a span of approximately 3 to 6 months, varying according to the species. They deposit batches of 3 to 40 transparent oval or spherical eggs in protective locations such as under leaves, in soil crevices, and other similar spots. Snails and slugs consume a diverse range of living plants and decaying plant material. They create irregular holes with smooth edges in leaves and blooms of plants, and are capable of trimming succulent plant sections. In addition, they have the ability to gnaw on both fruit and the bark of immature plants. Due to their preference for juicy leaves or blossoms, they mainly infest young plants and plants with soft stems.

An effective snail and slug management program depends on the utilization of many techniques. The initial measure involves minimizing, as much as feasible, any potential hiding spots for snails or slugs during daylight hours. Boards, stones, garbage, weedy regions surrounding tree trunks, leafy branches growing in proximity to the ground, and dense ground covers like ivy are optimal habitats for sheltering. There will exist shelters that cannot be eliminated, such as low ledges on fences, the undersides of wooden decks, and water meter boxes. Establish a consistent routine of capturing and eliminating snails and slugs in these specific locations.

Caterpillar

The bougainvillea looper is a caterpillar that measures approximately 1 inch in length and can be either green or brown in color. The organism is alternatively referred to as the inchworm or measuring worm. The looper larva exhibits excellent mimicry of stems and branches and primarily feeds during the nocturnal hours. This behavior explains why you may observe the damage caused by the larva but struggle to locate the actual culprit on the plant. Bacillus thuringiensis, available in various products, is highly efficient in controlling the larval stages of the looper. Bacillus thuringiensis (Bt), a bacterial formulation, induces a pathogenic condition in certain species of caterpillars while posing no threat to beneficial insects, birds, humans, or other organisms. Loopers cease feeding within a few hours after consuming a sprayed leaf and perish a few days thereafter. A comprehensive application of spray is necessary to effectively manage the tree. (Bt will also manage other caterpillars that are present throughout the application.) Bacillus thuringiensis (Bt) is most efficient in controlling larvae of fruit tree leaf roller when they are in their early stages, measuring less than 1/2 inch in length. Typically, many applications of Bt are necessary for optimal effectiveness. Caterpillars must consume the insecticide in order to be eradicated.

Use in Adornment

Bougainvillea bracts are commonly used in lei making. Clip clusters of bracts, preferably in the early morning, and transport them in a paper or cloth bag. Wash with a cold-water soak. Bracts can be wrapped in damp newspaper, placed in a plastic container or paper box, and stored in a refrigerator at 40°F for up to 14 days.

Bougainvillea bracts are also used in haku (in Guam, mwar mwar), traditional, lei-like headbands made of leaves and flowers.

Future Trends in Precision Farming for Horticulture

The future of precision farming in horticulture is marked by scope for a lot of opportunities especially in a country like India. Being driven by technology and the evolving needs of agriculture, with interest in the technological advancement in the field precision technologies have a lot of wonders to add to horticulture especially for younger, new-generation farmers. As we look ahead, several trends are shaping the course of precision farming, providing insights into the sensible changes that lie ahead for horticulturists.

In the coming years, an increased integration of Artificial Intelligence (AI) and Machine Learning (ML) is expected to refine decision-making processes. These technologies are expected to enhance the predictive capabilities of precision farming systems, allowing for more accurate assessments and proactive interventions. Additionally, the expansion of Internet of Things (IoT) and its varied applications are expected to deepen the connectivity between sensors, machinery, and decision support systems, facilitating seamless data exchange and real-time adjustments in horticultural practices.

Additionally, the future holds a promising series of advancements in autonomous machinery, streamlining tasks such as planting, harvesting, and monitoring. This automation not only reduces manual labor but also ensures precision and efficiency in day-to-day operations. Overall, the future trends in precision farming for horticulture align with practical innovations, emphasizing the integration of advanced technologies to enhance productivity, sustainability, and the overall effectiveness of agricultural practices.

Chapter 13

Mechanized Harvesting of Horticultural Produce

Introduction: An Overview

Mechanization plays a crucial part in ensuring the future success of fruit growers in developed nations. As the population grows, the demand for food, feed, and industrial raw materials from agriculture and horticulture also increases. The demand for sustainable mechanization is also increasing in order to achieve maximum production. Horticulture is a crucial sector for expanding the range of agricultural activities in India. India ranks as the second-largest producer of fruits and vegetables. The horticultural area in India is 25,870 thousand hectares, with a total production of 314,671 thousand MT (Anon, 2018-19). India is a major producer and exporter of spices. Horticulture crops will enhance the livelihoods of small-scale farmers in India during the period of commercial and high-value agriculture. The implementation of agricultural mechanization in Indian horticulture, namely for tasks such as transplanting, sowing, spraying, intercultural operations, fertilizer application, and harvesting, would lead to improvements in horticulture production, productivity, and quality.

Mechanized harvesting revolutionizes the landscape of horticulture, offering efficiency, precision, and scalability to the process of gathering fruits, vegetables, and other produce. In contrast to traditional hand-picking methods, which are labor-intensive and often time-consuming, mechanized harvesting employs a variety of specialized machinery and equipment to streamline the harvesting process. From automated fruit pickers that delicately pluck ripe apples from trees to robotic lettuce harvesters that navigate fields with precision, these advancements in technology represent a significant leap forward in agricultural productivity. The integration of mechanized harvesting not only enhances the speed and efficiency of crop collection but also addresses labor shortages, reduces reliance on manual labor, and minimizes post-harvest losses. As we delve into the realm of mechanized harvesting of horticultural produce, we uncover the ingenuity, innovation, and impact of these technologies on modern agricultural practices, paving the way for a more sustainable and productive future in horticulture.

Horticultural Tools and Apparatus

Horticulture, the art and science of cultivating plants, demands precision, care, and the right tools. From tending to delicate flowers to nurturing robust vegetables, horticulturalists rely on an array of specialized tools and apparatus designed to facilitate every aspect of plant care. Whether it's pruning shears for shaping shrubs, watering cans for gentle hydration, or soil pH meters for ensuring optimal growing conditions, these instruments are indispensable companions in the gardener's arsenal. In this exploration of horticultural tools and apparatus, we delve into the diverse array of equipment that aids in nurturing flourishing gardens and landscapes, highlighting their importance in the cultivation of green spaces and the beauty they bring to our surroundings. Horticultural tools encompass a range of implements such as grafting knife, budding knife, pruning knife, multi-purpose chopping knife, flower scissors, chain saw, and potting material mixer. Modern horticultural mechanization encompasses a range of techniques and processes for growing, production, working operations, technical procedures, soil management systems, orchard tractors, mulching machines, post hole diggers, fruit harvesting equipment, shakers, and picking machines. The tasks of harvesting, trimming, and spraying are currently predominantly carried out manually and require automation.

Process of Mechanized Harvesting

Mechanized harvesting stands as a pinnacle of agricultural innovation, representing the convergence of engineering prowess and agricultural science to revolutionize the way we gather horticultural produce. At its core, mechanized harvesting embodies a sophisticated orchestration of machinery, robotics, and automation meticulously designed to streamline the arduous task of crop collection. From the lush orchards of citrus fruits to the sprawling fields of leafy greens, the process of mechanized harvesting encompasses a myriad of intricacies tailored to the unique needs of each crop type. Through a symphony of cutting-edge technologies, including sensor-based systems, robotic arms, and GPS-guided machinery, mechanized harvesting transcends the limitations of traditional manual labor, offering unparalleled precision, efficiency, and

> *Mechanical harvesting can yield a remarkably high rate of product output. These machines can efficiently gather fruit and vegetables in significantly less time compared to manual harvesting methods. Mechanical harvesting systems alleviate managerial issues related to labourers. Minimize agricultural yield reduction by implementing timely harvesting practices.*

scalability. As we embark on an exploration of the process of mechanized harvesting, we embark on a journey into the heart of agricultural innovation, where science and engineering converge to redefine the landscape of horticulture and pave the way for a more sustainable and productive future in food production.

Mechanized harvesting is the utilization of machines rather than human labor to gather crops. The utilization of mechanized methods for harvesting crops is gaining popularity in horticulture due to several benefits it offers compared to human harvesting. Below are few prevalent instances of automated harvesting in horticulture:

Mechanical Fruit Harvesters

Limb shaker, canopy shaker, trunk shaking, air blasting, and robotic harvesting are various mechanical methods used for harvesting. Limb shakers are devices used for fruit harvesting. Harvesting can be done by utilizing limb shaker technology on citrus fruits, apricots, peaches, and cherries. The shaker was operated by a 2-stroke (spark ignition) engine. A centrifugal clutch, in conjunction with a gearbox, was utilized in tandem with a slider crank mechanism to transfer power to a limb via a boom and C-shaped clamp. The shakers can be remotely controlled by the operator using the handle on the shaker. The limb shakers (Figures 13.1a-b) are highly efficient in removing a significant proportion of fruit by applying long strokes to the branches at a low frequency. This procedure may result in the bark and limbs of the tree being harmed, and it can even lead to the removal of immature fruit. The efficiency of the limb shaker is mostly influenced by the cultivars and working conditions present at the orchard.

Figure 13.1a. Limb Shaker.

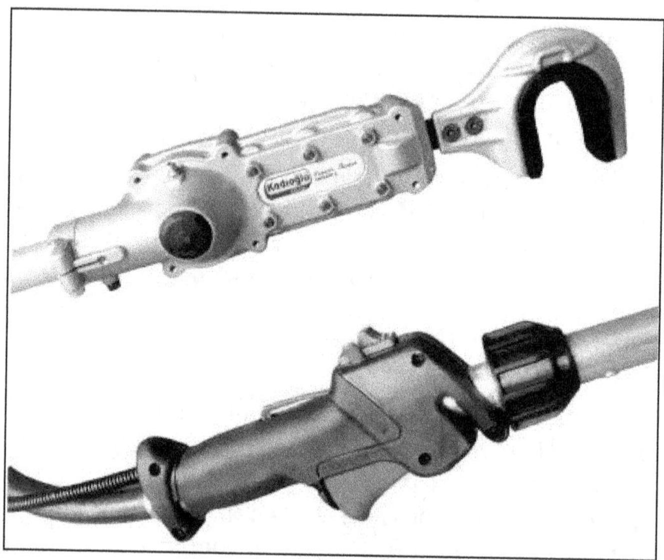

Figure 13.1b. Limb Shaker.

Canopy Shaker

Fruits are gathered utilizing a vibrating device that directly impacts the fruit or the branches that bear the fruit. The device can be securely fastened to additional branches and then moved up and down in a vertical motion to gather the fruit. There are two specific types of continuous canopy shakers that are expressly built for citrus fruit: a self-propelled unit and a tractor-drawn unit. The performance of this harvester is significantly influenced by the stroke and frequency of shaking. A canopy shaker is a mechanical harvesting method for oranges consisting of a vertical axis with 12 sets of free-floating tines that are 2 meters long and radiate from the vertical axis. The yield obtained during the harvest period is influenced by the frequency at which the tines shake, the depth at which the shaker is positioned within the crop canopy, and the force applied to detach the fruit (Figure 13.2).

Trunk Shaker

Deciduous fruits, olives, almonds, and citrus are primarily harvested with a trunk shaker. Typically, a tractor-mounted trunk shaker (Figure 13.3) is preferred over a hand-held shaker when it comes to shaking cultivars of various fruits. The tractor mounted shaker achieved a detachment rate of 72 per cent, which was higher than the 57 per cent detachment rate achieved by the hand-held shakers. High shaking frequency resulted in significant defoliation and bark damage.

Air Blast Harvester

The fruit was dislodged off the tree using a pneumatic burst of force. This technique utilizes oscillating air blast devices to increase the rate at which fruits are detached by adjusting the oscillation rate. The size of the tree structure, the weight of the fruit, and the fruit load of trees are all elements that impact the performance of an air blast harvester.

Figure 13.2. Canopy Shaker.

Figure 13.3. Trunk Shaker.

Fruit Harvesting

Mechanical fruit harvesting refers to the process of using machinery to collect fruits from orchards or fields. It typically involves specialized equipment designed to efficiently and gently detach fruits from trees or plants while minimizing damage to the produce. Mechanical harvesting offers several advantages over manual picking, including increased speed, reduced labor costs, and the ability to harvest large volumes of fruit in a shorter time frame. Various types of mechanical harvesters exist, each tailored to specific types of fruit and growing conditions. For example, there are machines designed for harvesting apples, oranges, grapes, berries, and more. These machines may use different mechanisms such as shaking, suction, or cutting to detach the fruits from their stems or branches. While mechanical harvesting can be highly efficient, there are also challenges associated with it. Ensuring that the machinery operates gently enough to avoid bruising or damaging the fruit is crucial, as damaged fruit may not be suitable for sale or processing. Additionally, some fruits, such as delicate berries, may require more specialized equipment to harvest effectively. Mechanical devices are commonly employed to gather fruit harvests such as apples, pears, and citrus fruits. Harvesting machines employ a shaking mechanism to dislodge the fruits from the tree, and subsequently gather them through the utilization of conveyor belts or other processes. The automated harvesting method is considerably swifter and more effective than manual harvesting, resulting in substantial reductions in labour expenses.

Mechanical fruit harvesting involves the use of specialized machinery to gather ripe fruits from trees or vines in orchards and fields. This process varies depending on the type of fruit being harvested and the specific conditions of the crop. Here's a detailed overview of mechanical fruit harvesting:

- ☆ **Preparation:** Before harvesting, farmers assess the readiness of the fruit by monitoring factors such as colour, size, sugar content, and firmness. Harvest timing is crucial to ensure optimal fruit quality and flavour.

- ☆ **Selection of Equipment:** Farmers select the appropriate harvesting machinery based on factors such as the type of fruit, orchard layout, and harvesting conditions. Different types of fruit may require specialized equipment, such as shakers, catch frames, conveyors, or picking arms.

- ☆ **Shaking:** For tree fruits like apples, pears, and cherries, shaking is a common method of mechanical harvesting. Harvesters equipped with shaking mechanisms gently shake the tree limbs or the entire tree, causing the ripe fruits to fall onto catching frames or conveyor belts positioned beneath the tree.

- ☆ **Catching and Conveyance:** As the fruits are dislodged from the trees, they fall onto catching frames or conveyor belts that collect and transport them to a central collection point or sorting area. Some harvesting machines incorporate vibration or suction systems to assist in dislodging fruits and guiding them onto conveyors.

☆ **Sorting and Grading:** Once the fruits are collected, they undergo sorting and grading to remove damaged or unripe fruits and ensure uniformity in size and quality. This may involve manual inspection or automated sorting systems equipped with cameras, sensors, and conveyor belts.

☆ **Packaging and Storage:** After sorting, the harvested fruits are packaged and prepared for storage or transportation to markets or processing facilities. Packaging may vary depending on the type of fruit and market requirements, with options ranging from bulk bins to individual consumer packaging.

☆ **Maintenance and Calibration:** Regular maintenance and calibration of harvesting machinery are essential to ensure optimal performance and minimize fruit damage. This includes inspecting and lubricating moving parts, adjusting shaking intensity, and monitoring conveyor speed.

☆ **Environmental Considerations:** Mechanical fruit harvesting may have environmental implications, such as energy consumption, emissions from machinery, and soil compaction. Farmers may implement sustainable practices to mitigate these impacts, such as using energy-efficient equipment, minimizing chemical inputs, and incorporating cover crops to improve soil health.

☆ **Integration with Other Practices:** Mechanical fruit harvesting is often integrated with other orchard management practices, such as pruning, pest control, and irrigation, to optimize overall orchard productivity and fruit quality.

Overall, mechanical fruit harvesting offers farmers a cost-effective and efficient way to harvest fruits while maintaining quality and maximizing yields. However, successful mechanical harvesting requires careful planning, proper equipment selection, and skilled operation to ensure optimal results.

Vegetable Harvesting

Automated harvesting is also employed for vegetable crops such as tomatoes, cucumbers, and peppers. Harvesting machines employ many techniques such as vibration, suction, and cutting to gather crops. These machines can be tailored to harvest various varieties of vegetables and to prevent crop damage during the harvesting process.

Mechanical vegetable harvesting involves the use of machinery to gather vegetables from fields or gardens. Similar to fruit harvesting, mechanical methods offer advantages like increased efficiency, reduced labor costs, and the ability to handle larger volumes of produce. Various types of machinery are employed for vegetable harvesting, depending on the type of vegetable being harvested and the specific conditions of the crop. Some common types of mechanical harvesters include:

☆ **Vegetable Harvesters:** These are typically large machines equipped with conveyor belts or picking arms designed to gather vegetables such as lettuce, spinach, cabbage, and other leafy greens. They may use cutting or pulling mechanisms to detach the vegetables from the plants while minimizing damage.

☆ **Root Crop Harvesters:** Machinery like potato harvesters are specialized for digging up root vegetables like potatoes, carrots, and beets. These machines usually have digging blades or belts that lift the vegetables out of the ground, followed by mechanisms to separate the vegetables from soil and debris.

☆ **Bean and Pea Harvesters:** Mechanical harvesters for legumes like beans and peas often utilize mechanisms to strip the pods from the plants. These machines may employ vibration, rollers, or brushes to detach the pods without damaging them.

☆ **Tomato Harvesters:** Tomato harvesters are designed to pick ripe tomatoes from the vines. They may use shaking or suction mechanisms to detach the tomatoes from the plants, followed by conveyors or sorting systems to separate them based on size and quality.

☆ **Sweet Corn Harvesters:** Mechanical harvesters for sweet corn typically involve machines that strip the ears from the stalks and remove the husks. They may use cutting blades or snapping mechanisms to detach the ears while leaving the stalks intact.

Mechanical vegetable harvesting can significantly improve efficiency in large-scale agriculture, allowing farmers to harvest their crops more quickly and cost-effectively. However, like fruit harvesting, it requires careful calibration and operation to ensure that the machinery handles the vegetables gently to minimize damage and maintain quality.

Grain Harvesting

Automated harvesting is also prevalent in cereal crops such as wheat, corn, and rice. Combine harvesters, also known as harvesting machines, are utilized to chop and thresh the grain, separating the seeds from the stalks and husks. These machines possess a great level of efficiency, enabling them to gather hundreds of acres of grain within a single day.

Mechanical grain harvesting refers to the process of using machinery to gather mature grain crops from fields. This method has largely replaced manual harvesting techniques due to its efficiency and ability to handle large-scale agricultural operations. The primary machine used for mechanical grain harvesting is the combine harvester, often simply called a combine. Combines are versatile machines capable of performing several functions in one pass through the field:

- ☆ **Cutting:** Combines are equipped with a cutting mechanism, typically a rotating blade or header, which cuts the mature grain plants at ground level.

- ☆ **Threshing:** Once the grain plants are cut, the combine separates the grain kernels from the rest of the plant through a process called threshing. Threshing mechanisms inside the combine separate the grain from the straw or chaff.

- ☆ **Separating:** After threshing, the combine separates the grain from the straw, husks, and other plant material. This is usually achieved using a series of sieves or screens that allow the grain to fall through while retaining the remaining material.

- ☆ **Cleaning:** The separated grain is then cleaned to remove any remaining debris or impurities, such as small pieces of straw or chaff.

- ☆ **Grain Collection:** Finally, the clean grain is collected in a grain tank onboard the combine. Once the tank is full, the grain can be unloaded into a truck or storage facility for further processing or sale.

Combines come in various sizes and configurations to suit different types of grain crops and field conditions. They may also be equipped with advanced technologies such as GPS guidance systems and yield monitors to optimize harvesting efficiency and collect data on crop yields. Mechanical grain harvesting offers several advantages over manual harvesting, including increased speed, reduced labor costs, and the ability to harvest crops in a timely manner, even in large fields. However, proper maintenance and operation of the equipment are essential to ensure optimal performance and grain quality.

Flower Harvesting

Flower crops such as roses and carnations also utilize mechanized harvesting. Harvesting machines employ blades or alternative cutting devices to sever the blooms at the lowermost part of the stem. These robots have the capability to gather a large number of flowers every hour, which leads to a decrease in labour expenses and an improvement in productivity.

Mechanical flower harvesting involves the use of specialized machinery to gather flowers from fields or greenhouses. While flower harvesting is often done manually due to the delicate nature of flowers and the need to preserve their quality, mechanical methods have been developed to streamline the process, particularly for certain types of flowers grown on a larger scale.

Some common types of machinery used for mechanical flower harvesting include:

- ☆ **Flower Harvesting Machines:** These machines are designed to gently and efficiently cut flowers from their stems. They may use rotating blades or cutting mechanisms that can be adjusted to the specific height required for each type of flower.

☆ **Bunching Machines:** After flowers are harvested, bunching machines can be used to gather and bunch them into marketable bouquets or bunches. These machines typically use conveyor belts or mechanical arms to gather and arrange the flowers into uniform bunches.

☆ **Sorting and Grading Machines:** Once flowers are harvested and bunched, sorting and grading machines can be used to inspect and classify them based on factors such as size, color, and quality. These machines help ensure consistency and uniformity in the final product.

☆ **Packaging Machines:** Finally, packaging machines can be used to wrap or package the flowers for transportation and sale. These machines may use various packaging materials such as sleeves, wraps, or boxes to protect the flowers during transit.

While mechanical flower harvesting offers benefits such as increased efficiency and reduced labor costs, it also presents challenges, particularly in handling delicate flowers without causing damage or bruising. As a result, mechanical harvesting methods are often used in conjunction with manual harvesting techniques, with machinery used for bulk harvesting and manual labour employed for more delicate or specialty flowers. Overall, mechanical flower harvesting can be a valuable tool for commercial flower growers looking to increase efficiency and productivity in their operations, particularly for flowers grown on a larger scale for commercial markets.

Advantages of Mechanized Harvesting of Horticultural Produce

Mechanized harvesting of horticultural produce has revolutionized agricultural practices, offering a myriad of advantages that enhance efficiency, productivity, and overall farm operations. By harnessing the power of specialized machinery, farmers can streamline the harvesting process, reduce labor costs, and optimize crop yields. This modern approach to harvesting not only improves the bottom line for farmers but also ensures the timely and efficient delivery of high-quality produce to markets. In this era of evolving agricultural technology, mechanized harvesting stands as a cornerstone of modern farming practices, facilitating sustainable and profitable agriculture worldwide. Mechanized harvesting of horticultural produce offers several advantages over manual harvesting methods:

☆ **Increased Efficiency:** Mechanized harvesting enables farmers to harvest larger areas of crops in a shorter amount of time compared to manual harvesting. This efficiency can lead to higher overall productivity and reduced labour costs.

☆ **Labor Savings:** Mechanized harvesting reduces the need for manual labour, which can be costly and time-consuming to procure, especially during peak harvesting seasons when labour shortages may occur. By relying on machinery, farmers can mitigate these challenges and ensure timely harvests.

☆ **Consistency and Quality Control:** Machinery can be programmed to harvest crops with consistent precision, leading to uniformity in produce size, shape, and quality. This consistency is important for meeting market standards and consumer expectations, ultimately enhancing the marketability of the harvested produce.

☆ **Reduced Physical Strain:** Harvesting can be physically demanding work, especially for crops that require bending or repetitive motion. Mechanized harvesting reduces the physical strain on workers, leading to improved worker safety and reduced risk of injuries.

☆ **Extended Harvest Windows:** Machinery can operate in various weather conditions and can often harvest crops during periods when manual labour may not be feasible, such as during inclement weather or when labour shortages occur.

☆ **Optimized Harvest Timing:** Mechanized harvesting allows farmers to optimize harvest timing based on factors such as crop maturity, weather conditions, and market demand. This flexibility can help farmers maximize yield and quality while minimizing post-harvest losses.

☆ **Cost Savings:** While the initial investment in harvesting machinery can be significant, mechanized harvesting often leads to long-term cost savings through increased efficiency, reduced labour costs, and improved productivity.

☆ **Scalability:** Mechanized harvesting is highly scalable, allowing farmers to expand their operations without proportionally increasing labour costs. This scalability is particularly advantageous for large-scale commercial horticultural operations.

☆ **Data Collection and Analysis:** Some modern harvesting machinery is equipped with sensors and data collection capabilities, allowing farmers to gather valuable information about crop yields, performance, and field conditions. This data can inform decision-making processes and help optimize farming practices for increased efficiency and profitability.

Overall, mechanized harvesting of horticultural produce offers numerous benefits that can help farmers improve productivity, reduce costs, and enhance the overall sustainability of their operations.

Disadvantages of Mechanized Harvesting of Horticultural Produce

While mechanized harvesting of horticultural produce brings significant benefits to modern agriculture, it is not without its drawbacks. As farms increasingly rely on specialized machinery to streamline operations, there are concerns regarding the initial investment costs, dependence on technology, and potential environmental impacts. Understanding these disadvantages is crucial for farmers and stakeholders as they navigate the complexities of adopting mechanized harvesting practices. Despite its undeniable advantages, the mechanization of harvesting presents challenges that require careful consideration and management to ensure sustainable and responsible agricultural practices.

While mechanized harvesting of horticultural produce offers several advantages, it also comes with some disadvantages:

☆ **High Initial Investment:** The purchase and maintenance costs of harvesting machinery can be substantial, especially for small-scale or beginning farmers. This initial investment may pose a barrier to entry for some farmers or require significant capital outlay.

☆ **Dependence on Machinery:** Mechanized harvesting relies on the proper functioning of machinery, which can break down or require repairs. Dependence on machinery means that any mechanical failures or downtime can disrupt harvesting schedules and potentially lead to losses in crop yield and quality.

☆ **Complexity and Maintenance:** Harvesting machinery can be complex to operate and maintain, requiring specialized knowledge and skills. Farmers may need to invest time and resources in training operators and ensuring regular maintenance to keep machinery in optimal working condition.

☆ **Limited Suitability for Certain Crops:** While mechanized harvesting is well-suited for many horticultural crops, it may not be suitable for all crops or varieties. Some crops may be too delicate or have specific harvesting requirements that are difficult to meet with machinery, leading to potential damage or loss of quality.

☆ **Impact on Soil and Plants:** Heavy machinery used for mechanized harvesting can compact soil and cause damage to crops if not operated carefully. Soil compaction can impair soil structure and drainage, leading to decreased soil fertility and increased erosion risk.

☆ **Loss of Employment:** Mechanized harvesting reduces the need for manual labor, which can result in job displacement for agricultural workers who rely on harvesting jobs for income. This can have socio-economic implications for rural communities dependent on agricultural employment.

☆ **Environmental Concerns:** The use of machinery in harvesting can contribute to environmental issues such as fuel consumption, air pollution, and greenhouse gas emissions. Additionally, the disposal of agricultural machinery at the end of its lifecycle can pose challenges for waste management and environmental sustainability.

☆ **Limited Adaptability:** Harvesting machinery may have limitations in adapting to diverse field conditions, crop varieties, or terrain types. This lack of adaptability can restrict the applicability of mechanized harvesting in certain regions or farming systems.

☆ **Loss of Traditional Knowledge:** Mechanized harvesting may lead to a loss of traditional farming practices and knowledge associated with manual harvesting methods. This loss of cultural heritage and traditional farming wisdom can have intangible impacts on farming communities and agricultural landscapes.

Overall, while mechanized harvesting offers many benefits, it is important for farmers to carefully consider the potential disadvantages and weigh them against the advantages when making decisions about adopting mechanized harvesting practices.

Chapter 14

Post-Harvest Management of Horticultural Produce for Export

Introduction

Agriculture plays a crucial role in the Indian Economy, and this reliance is expected to persist in the foreseeable future. In 2013-14, the Agriculture and Allied Sector accounted for around 13.9 per cent of India's GDP, as reported by the Ministry of Agriculture in 2015. Additionally, this sector employs approximately 50 per cent of the country's workers. India is a highly fertile country that cultivates a diverse range of fruits, vegetables, spices, ornamental plants, and medicinal plants. India is the second-largest global producer of veggies. Advancements in production technologies have led to a steady growth in food production in India.

The post-harvest management, processing, and value addition of fruits, vegetables and ornamental crops are currently underdeveloped in the country. It is crucial to closely examine some fundamental aspects of post-harvest management of horticultural produce to reduce wastage and increase profits for farmers. This would contribute to the augmentation of per capita availability, enhance the economic situation of the farmers, and guarantee equitable distribution of agricultural goods across the nation. Value addition is a highly effective method for minimizing post-harvest losses in flower crops. India's vast genetic diversity allows for the production of a diverse range of value-added products, such as dry flower pigments, neutraceutical compounds, essential oils, petal embedded papers, and floral crafts. These products are targeted for both domestic and international markets.

However, inadequate post-harvest management, processing, value addition, and storage practices have resulted in significant losses in agricultural goods. Regrettably, the substantial scale of production results in a significant yearly post-harvest loss of vegetables, estimated to be between 10-25 per cent (Selvakumar 2014). This loss is primarily attributed to inadequate post-harvest management procedures.

The World Bank Report states that India experiences post-harvest losses of 12 to 16 million metric tonnes of food grains annually. This quantity of food could potentially provide sustenance for one-third of India's impoverished population, as determined by the World Bank (Chaturvedi and Raj, 2015). The financial magnitude of these losses exceeds Rs 50,000 crores annually (Singh, 2010). Annually, India produces around 263.2 million metric tons of food grains, as of 2013-14 (Chaturvedi and Raj, 2015). Of this amount, farmers store 60-70 per cent for their personal consumption.

The optimal post-harvest treatment of horticultural goods varies depending on the specific commodity. In order to maximize the post-harvest shelf life and quality of a product, it is crucial for growers, wholesalers, exporters, and retailers to have a thorough understanding of its individual requirements. Hence, it is imperative to minimize post-harvest losses in order to sustainably provide for the increasing population of the country. The processing of food products, primarily within the context of a cottage industry, has been a deeply ingrained traditional practice in various ethnicities throughout the country. Nevertheless, due to shifting lifestyle patterns, rising income levels, and a growing inclination towards ready-to-eat and packaged goods, the Food Processing Industry has experienced a substantial growth in importance. India benefits from a diverse set of favorable agro-climatic conditions that allow for the cultivation of a wide variety of flowers, potted plants, leaves, and scented flowers across the country, practically year-round in different regions. The floriculture sector in India is distinguished by the cultivation of traditional flowers (loose flowers) and cut flowers in both open field and protected environment conditions, respectively. India also boasts a robust dried flower business, making a significant contribution to the country's overall trade. Additional sectors such as fillers, potted plants, seeds and planting material, the turf grass business, and value-added products also make a significant contribution to the total expansion of the floriculture sector. Loose flowers are mostly utilized for religious rituals, crafting garlands and other floral decorations, extracting essential oils, pigments, and dyes. Harvest flowers by cutting the stem or branch, ensuring that the desired length includes the blossom, buds, and leaves, all with a specific purpose in mind. The cut flowers are commonly utilized indoors in vases to decorate houses, drawing rooms, businesses, hotels, and other spaces. They are also frequently arranged into appealing bouquets. Post-harvest losses refer to the losses that occur from the time of harvest till the produce is received by consumers. It may measure losses both in terms of amount and quality. Post-harvest losses are more economically and physically burdensome compared to pre-harvest losses, both in terms of financial costs and labour requirements. Vegetables are extremely prone to spoilage due to their high moisture content, which often ranges from 80 per cent to 90 per cent. Live commodities maintain their life functions, such as respiration and transpiration, even after being harvested. When the fruit is connected to the parent plant, it receives water and photosynthates. However, losses incurred during the post-harvest stage are not replenished, causing the produce to rely only on its own food reserves and moisture content. Consequently, they deteriorate rapidly. Water is lost from the product by transpiration, while the food store is depleted through respiration. Water loss, also known as transpiration, significantly impacts the quality of vegetables. Loss of water can have several detrimental effects on the quality of a product, such as wilting, shriveling, flaccidness, soft texture, and diminished nutritional content. Additionally, it can also result in a decrease in saleable weight. The rate of water loss and its influence will differ depending on the product, as seen in Tables 14.1 and 14.2. As an illustration, the allowable number of losses can vary from 3 per cent for lettuce to 10 per cent for onions. Products range in their capacity for water loss due to variations in morphological characteristics, such as the thickness and composition of the cuticle, as well as the presence or absence of stomata and lenticels. Stomata and lenticels are structures that enable the movement of gasses and moisture in and out of the plant. The variations in these products are influenced by the stage of development. Furthermore, there are morphological distinctions among different sorts of items. Respiration produces water as a byproduct. To minimize water loss, it is advisable to chill products, maintain a high relative humidity in the storage environment, regulate air circulation, and, if allowed, apply surface coatings or plastic film (Chris and Jacqueline, 2012). The duration of the cut flowers' potential usefulness is determined by various events that occur before, during, and after the harvesting process. Approximately 70 per cent of the enduring characteristics of cut flowers are mostly determined at the time of harvest, whereas post-harvest influences account for 30 per cent of the resulting effects. The post-harvest behavior and longevity of flower species and cultivars exhibit significant variation. Variances in blossom duration and excellence across different types may arise from differences in their anatomical, physiological, physical, biochemical, and genetic composition. The duration of cut flower life is influenced by factors such as the amount of carbohydrates stored in the flowers, the osmotic concentration and pressure potential of the cells in the petals, the functioning of the stomata, variations in the diameter and rigidity of the stem, and the presence of different bacterial and fungal species in the water in the vase. Value addition occurs at each stage of the floriculture trade to provide a unique product that meets the needs of diverse consumers. The trends in floriculture closely align with those in the fashion business, resulting in a significantly faster evolution of the floriculture sector compared to other industries. Various value-added products are derived from flower crops, including rose water, rose oil, otto or attar, concrete, absolute of rose, and *itra* of rose. Transpiration losses for vegetables stored at various relative humidity and Post-harvest losses in major vegetables are summarized in Tables 14.1 and 14.2 respectively.

Table 14.1. Transpiration Losses for Vegetables Stored at Various Relative Humidity (Chris and Jacqueline, 2012)

Crops	Storage Temperature (°F)	Percentage Weight Loss per Day			
		95 per cent RH	90 per cent RH	85 per cent RH	80 per cent RH
Carrots	32	0.315	0.630	0.945	1.260
Cabbage	32	0.058	0.116	0.175	0.233
Celery	32	0.460	0.920	1.380	1.840
Lettuce	32	1.930	3.860	5.790	7.730
Potatoes	45	0.070	0.141	0.211	0.282
Tomatoes	45	0.060	0.119	0.180	0.240

Table 14.2. Post-harvest Losses in Major Vegetables (Selvakumar 2014)

Name of Vegetable	Post-harvest Losses as Percentage of Production
Beans and peas	7-12
Brinjal	10-13
Cabbage	7-15
Cauliflower	10-15
Garlic	1-3
Onion	15-30
Potato	15-20
Tomato	10-20

Causes of Post-harvest Losses in Fruits and Vegetables

Post-harvest losses in fruits and vegetables represent a significant challenge in the agricultural sector worldwide, posing economic, nutritional, and environmental implications. These losses occur at various stages along the supply chain, from harvesting to consumption, and can be attributed to a multitude of factors. Understanding the complex interplay of these factors is crucial for devising effective strategies to mitigate post-harvest losses and ensure food security. Environmental conditions, such as temperature, humidity, and light exposure, play a pivotal role in determining the shelf life of fruits and vegetables. Additionally, factors such as improper handling, inadequate storage facilities, transportation challenges, and market dynamics further contribute to post-harvest losses. Addressing these challenges demands a comprehensive approach that integrates technological innovations, infrastructure development, policy interventions, and behavioural changes among stakeholders. By addressing the root causes of post-harvest losses, we can not only reduce food waste but also enhance food availability, improve livelihoods, and promote sustainable agricultural practices.

The factors contributing to post-harvest losses differ significantly among locations and become increasingly intricate. The subsequent factors contributing to post-harvest losses are as follows:

- ☆ Wilting or shrinkage occurs due to moisture loss.
- ☆ Loss of photosynthates, such as carbohydrates and proteins, takes place.
- ☆ Physical damage can be caused by pests and diseases.
- ☆ Physiological changes lead to a decline in quality.
- ☆ Fiber development occurs.
- ☆ Potatoes may undergo greening.
- ☆ Microbial causes include insects and rodents.
- ☆ Enzymatic activity of the plant or food can contribute to deterioration.
- ☆ Chemical reactions that are not catalyzed by tissue enzymes can occur.
- ☆ Physical changes, such as freezing, bruising, drying, and pressure, can also occur.

Major Factors Affecting Post-Harvest Life of Horticultural Crops

The post-harvest life of horticultural crops, encompassing fruits, vegetables, and ornamental plants, is influenced by a myriad of factors that collectively determine their quality, shelf life, and marketability. Understanding these factors is essential for ensuring optimal post-harvest management practices, minimizing losses, and maximizing economic returns for growers, distributors, and consumers alike. Environmental conditions, such as temperature, humidity, and light exposure, significantly impact the physiological processes and biochemical reactions occurring in harvested crops, directly affecting their shelf life and quality. Furthermore, factors related to pre-harvest management, including cultivation practices, irrigation regimes, and pest and disease control, can influence the susceptibility of horticultural crops to post-harvest deterioration. Post-harvest handling practices, such as harvesting methods, sorting, grading, packaging, and transportation, also play a crucial role in determining the extent of losses and quality degradation during storage and distribution. Additionally, advancements in post-harvest technologies, including cold chain infrastructure, controlled atmosphere storage, and modified atmosphere packaging, offer opportunities to extend the shelf life of horticultural crops and minimize post-harvest losses. By comprehensively understanding and addressing these major factors affecting post-harvest life, stakeholders can enhance the efficiency, sustainability, and resilience of horticultural supply chains, thereby ensuring food security, promoting economic development, and reducing food waste.

There are two methods for decreasing the number of veggies lost after harvesting. One effective method to reduce loss is to implement a scientific approach to post-harvest treatment of vegetables. An alternative method for minimizing losses during the transformation into value-added goods. The post-harvest technology of vegetable crops focuses on the development of suitable methods to minimize post-harvest losses, prevent spoiling, and maximize the utilization of products in a nutritious and safe manner.

Practices for Managing Post-harvest Losses

Managing post-harvest losses in horticultural produce is paramount for ensuring food security, sustainability, and economic viability within the agricultural sector. The journey from farm to table involves numerous stages where losses can occur, including harvesting, handling, storage, transportation, and marketing. Implementing effective management practices at each stage is essential for minimizing these losses and maximizing the value of horticultural produce. From adopting proper harvesting techniques and optimizing storage conditions to investing in infrastructure and technologies for transportation and distribution, a holistic approach is required. Furthermore, incorporating innovative solutions such as cold chain management, controlled atmosphere storage, and packaging technologies can significantly extend the shelf life of horticultural produce and reduce post-harvest losses. By prioritizing these practices, stakeholders across the supply chain can enhance food availability, improve livelihoods, and promote sustainable agricultural systems. Post-harvest losses can be minimized by implementing breeding strategies to enhance shelf life, optimizing pre-harvest factors and harvesting techniques, employing effective ways for handling, marketing, packaging, shipping, and storage, and developing suitable processing technology.

Selection of Varieties

It is important to breed and choose distinct varieties of vegetables that have improved storage and processing qualities, as well as reduced susceptibility to damage during handling. Some examples of cultivars with extended shelf life include Arka Vishal, Pusa Gaurav (Tomato), Arka Nidhi, and Arka Neelakandh (Brinjal).

Harvesting at Proper Stage

Harvesting should be conducted at the appropriate time to minimize damage and loss, while maximizing efficiency and minimizing expenses. Harvesting should be conducted during the early morning or late nighttime hours. Avoid temperatures exceeding 27°C throughout the harvesting process. The products destined for faraway markets are picked in the evening and transported during the cooler nighttime hours, whereas commodities intended for local markets are harvested in the early morning. Harvesting should be postponed following rainfall or irrigation. Harvesting at the ideal stage of maturity guarantees the highest level of quality and production. It is important to use caution in order to prevent any physical harm to the product.

Sorting/Grading

The process of sorting harvested vegetable produce involves the removal of vegetables that are diseased, damaged, deformed, over mature, pest attacked, or rotten. To prevent the transmission of infection to healthy vegetable and fruit products, any plants affected by disease or insects should be removed. Implementing a methodical grading system, along with suitable packing and storage techniques, will prolong the shelf life, maintain the quality and freshness, and enhance the overall wholesomeness of the produce. Additionally, it will significantly minimize losses and reduce marketing expenses. Horticultural produce must undergo sorting and grading based on factors including maturity, size, shape, colour, weight, absence of insects and pests, pesticide residues, and ripeness. Onion, potato, tomato, chillies, okra, and French beans are categorized based on their size, shape, weight, and level of maturity. Preventing the dissemination of illnesses is achieved by removing horticultural crops that are of inferior quality or affected by diseases. Horticultural crops are typically evaluated and categorized based on their dimensions and mass (Nath, 2013).

Washing/Cleaning of Produce

The produce undergoes a cleaning process to eliminate any dirt, dust, insects, mold, and spray residues, as well as to enhance its visual appeal. Onion, garlic, okra, and mushrooms are not post-harvest washed. For surface disinfection, you can use a chemically gentle detergent (soap solution), glacial acetic acid, or a 1 per cent NaCl solution. Water that has been treated with chlorine, specifically at a concentration of 100 parts per million, is also efficient in removing contaminants from surfaces. Prior to packaging, it is necessary to rinse fruits and vegetables once more with clean water and allow any extra water to evaporate.

Trimming

Trimming of horticultural crops is a critical post-harvest practice aimed at enhancing the quality, appearance, and marketability of harvested produce. This process involves the removal of unwanted or damaged parts, such as leaves, stems, and roots, to improve aesthetics, reduce spoilage, and prolong shelf life. Trimming not only enhances the visual appeal of horticultural crops but also helps minimize the risk of microbial growth and physiological disorders during storage and transportation. Proper trimming techniques, coupled with efficient handling and packaging, play a vital role in maintaining the freshness and integrity of the produce, thereby maximizing its value and minimizing post-harvest losses. As an integral part of post-harvest management, trimming contributes to ensuring that consumers receive high-quality, nutritious, and visually appealing horticultural products, thus facilitating sustainable agricultural practices and promoting market competitiveness. Trimming is a horticultural practice commonly employed in crops such as cabbage and lettuce. To eliminate undesirable, discoloured, decayed, and impaired sections. Trimming improves the visual appearance, minimizes the decay of the produce, and makes it easier to handle, package, and transport.

Curing

Curing is a crucial post-harvest practice in horticulture that involves the controlled exposure of harvested crops to specific environmental conditions to enhance their quality, flavor, and storability. This process is particularly important for certain horticultural crops such as root vegetables, tubers, and bulbs, where curing helps initiate physiological changes that improve their taste, texture, and resistance to decay. During curing, harvested crops are typically subjected to elevated temperatures and high humidity levels for a specific duration, allowing them to undergo biochemical reactions such as starch to sugar conversion, lignification, and cell wall strengthening. These changes not only enhance the flavor and nutritional content of the produce but also reduce susceptibility to pathogens and physical damage during storage and transportation. Curing also plays a significant role in the development of desirable traits in certain crops. For example, in crops like onions and garlic, curing promotes the formation of a protective dry outer skin layer, which helps extend their shelf life and enhances their marketability. Similarly, in root vegetables like sweet potatoes, curing triggers the healing of cuts and bruises incurred during harvest, leading to improved wound healing and reduced post-harvest losses.

Proper implementation of curing practices requires careful monitoring of environmental parameters such as temperature, humidity, and airflow, as well as adherence to specific curing protocols tailored to the requirements of different crops. Additionally, advancements in curing technologies, such as controlled atmosphere curing chambers and humidity-controlled storage facilities, have further optimized the curing process, allowing for precise control over the curing conditions and improved quality retention of horticultural crops. Overall, curing is an essential post-harvest practice in horticulture that not only enhances the quality and storability of harvested crops but also contributes to reducing post-harvest losses, improving market competitiveness, and ensuring a sustainable supply of nutritious and flavourful produce to consumers. Curing is a method used to reinforce and protect the outer layer of roots and tubers in crops. This process involves subjecting the crops to specific temperature and humidity conditions for a set period of time. By doing so, a corky layer is formed, which helps to extend the shelf life of these crops. This layer acts as a barrier, preventing water loss and protecting against decay-causing organisms. In the cultivation of bulb crops such as onions and garlic. Curing is a method of drying that results in the hardening of the outer skin and the tightness of necks. The optimal conditions for potato curing are a temperature of approximately 200 degrees Celsius and a relative humidity of 80 per cent.

Waxing

Waxing is a key post-harvest treatment used in horticulture to extend the shelf life and enhance the appearance of fruits and vegetables. This process involves the application of a thin layer of wax onto the surface of the produce, forming a protective barrier that helps retain moisture, reduce dehydration, and minimize physiological deterioration during storage and transportation. The primary purpose of waxing in horticulture is to preserve the quality and freshness of harvested crops by slowing down the loss of moisture and the onset of decay processes. By forming a hydrophobic layer on the surface of the produce, waxing helps prevent water loss, thereby maintaining turgidity and crispness in fruits and vegetables. Additionally, waxing can help reduce the incidence of superficial damage such as bruising, scuffing, and abrasions, which can occur during handling and transportation. Moreover, waxing enhances the visual appeal of horticultural produce by giving it a glossy appearance, which can make the fruits and vegetables more attractive to consumers. This aesthetic improvement can be particularly beneficial for increasing marketability and consumer acceptance, thereby improving sales and reducing waste.

Different types of waxes, including natural waxes derived from plants such as carnauba and synthetic waxes, are used in horticulture based on factors such as the type of produce, desired shelf life, and regulatory requirements. The application of wax is typically done using specialized equipment that ensures uniform coverage and adherence to quality standards. While waxing offers numerous benefits in preserving and enhancing the quality of horticultural produce, it is essential to adhere to proper application procedures and regulatory guidelines to ensure food safety and consumer satisfaction. Additionally, ongoing research and technological advancements in wax formulations and application techniques continue to improve the efficacy and sustainability of waxing practices in horticulture, contributing to the overall efficiency and competitiveness of the industry.

Waxing is primarily performed to limit transpiration and decrease desiccation, hence improving the longevity of storage. The wax closes the stem at the petiole and the pores on the surface of fruits, which are the primary pathways for transpiration. The process of wax accumulation on the surface of fruits or vegetables is a significant pathway for transpiration. Waxing enhances the visual appeal of vegetables. Paraffin wax, Carnauba wax, and different resins are commonly utilized for the formulation of wax emulsions. Waxes are commonly administered using foaming, spraying, and brushing techniques, with foaming being the most effective method due to its ability to create a very thin covering. Common coating materials include semperfresh, extend, and waxol. Vegetables like tomato, brinjal, sweet pepper, cucumber, muskmelon, carrot, *etc.* are commonly coated with a water emulsion through dipping or spraying. This is done to slow down the loss of moisture from the vegetables and enhance their shine. This approach of maintaining the product's integrity and shine is not commonly followed in our country. The shelf life of fruits and vegetables can be prolonged by using various plant-based coating materials such as neem extract, tulsi extract, aloe vera extract, and so on. The local application of (Nath *et al.*, 2013) can be utilized at the farmer level to harness its antifungal effects, hence increasing the profitability of agricultural yields.

Precooling

Precooling is a vital post-harvest practice in horticulture aimed at rapidly lowering the temperature of harvested crops to their optimal storage temperature. This process is essential for preserving the quality, freshness, and shelf life of horticultural crops by slowing down metabolic processes, reducing respiration rates, and inhibiting the growth of spoilage microorganisms. The primary objective of precooling in horticulture is to remove field heat from freshly harvested crops as quickly as possible to prevent heat-related deterioration and ensure uniform cooling throughout the produce. Field heat, accumulated during harvesting and handling activities, can accelerate physiological processes such as ripening, softening, and decay, leading to reduced quality and increased susceptibility to post-harvest losses.

Various methods are used for precooling horticultural crops, including forced-air cooling, hydrocooling, vacuum cooling, and evaporative cooling, each tailored to the specific requirements of different crops and production systems. Forced-air cooling involves circulating chilled air around the produce to extract heat, while hydrocooling immerses the crops in cold water to facilitate rapid heat transfer. Vacuum cooling utilizes reduced pressure to lower the boiling point of water, thereby evaporating moisture from the produce and cooling it rapidly. Evaporative cooling involves the evaporation of water from the surface of the produce, leading to a cooling effect through latent heat exchange. Proper precooling practices not only help extend the shelf life of horticultural crops but also maintain their quality attributes such as color, texture, flavor, and nutritional content. Additionally, precooling reduces the risk of moisture loss, weight loss, and physical damage during storage and transportation, thereby minimizing post-harvest losses and ensuring higher market value for the produce.

Efficient precooling requires careful attention to factors such as crop characteristics, cooling method, temperature management, airflow distribution, and handling practices. Advances in precooling technologies and equipment, coupled with improved logistics and infrastructure, have facilitated the widespread adoption of precooling practices in horticulture, contributing to enhanced product quality, market competitiveness, and sustainability across the supply chain. Pre-cooling refers to the procedure of eliminating the heat acquired by the harvested commodity, especially when it is harvested in hot weather. Pre-cooling reduces the rate of transpiration and respiration, slows down ripening, and reduces the strain on the cooling system of transportation or storage chambers. There are multiple techniques for the pre-cooling process, such as:

☆ Room cooling

☆ Hydro-cooling

☆ Contact icing

☆ Vaccum cooling

Post-Harvest Disease Control

Post-harvest disease control is a critical aspect of horticultural management, aiming to minimize losses and maintain the quality of harvested crops during storage, transportation, and marketing. Horticultural produce is particularly susceptible to various microbial pathogens, including fungi, bacteria, and molds, which can cause rot, decay, and spoilage if left unmanaged. Effective disease control strategies in post-harvest handling involve a combination of preventive measures, such as proper sanitation, temperature management, and humidity control, as well as curative treatments, such as fungicides, sanitizers, and biocontrol agents.

Preventive measures play a crucial role in minimizing the incidence and severity of post-harvest diseases by creating unfavourable conditions for pathogen growth and development. Sanitation practices, including the cleaning and disinfection of storage facilities, equipment, and packaging materials, help reduce the initial inoculum of pathogens and prevent their spread. Temperature management is also essential, as maintaining optimal storage temperatures can slow down the growth of pathogens and prolong the shelf life of horticultural produce. Additionally, controlling humidity levels and improving airflow within storage facilities can help reduce moisture buildup, which is conducive to fungal growth and disease development. In cases where preventive measures are insufficient, curative treatments may be employed to control post-harvest diseases and salvage affected crops. Chemical fungicides are commonly used to suppress fungal growth and prevent the spread of diseases, although their use may be restricted due to regulatory concerns and consumer preferences. Alternatively, sanitizers such as chlorine dioxide and hydrogen peroxide can be used to disinfect surfaces and reduce microbial contamination without leaving harmful residues. Biological control agents, including antagonistic microorganisms and microbial metabolites, offer eco-friendly alternatives for managing post-harvest diseases by outcompeting pathogens or inhibiting their growth through various mechanisms.

Integrated disease management approaches that combine multiple control strategies are increasingly being adopted in horticulture to enhance efficacy, reduce reliance on chemical inputs, and minimize environmental impacts. By implementing comprehensive disease control measures in post-harvest handling practices, horticultural stakeholders can mitigate losses, maintain product quality, and ensure the delivery of safe and healthy produce to consumers, thereby supporting sustainable agricultural systems and food security. Vegetables experience substantial harm as a result of the infiltration of fungus and bacteria, leading to the development of diseases and inflicting major losses after harvest. The high moisture content of plants makes them susceptible to microbial illness. Mechanical injuries, contamination from diseased vegetables, exposure to heat, and other environmental factors increase the susceptibility of products to illnesses. Fungicides can be used as sprays or dips, integrated in wax, or impregnated in packaging materials to suppress post-harvest infections.

Sprout Inhibition

Sprout inhibition is a crucial post-harvest practice in horticulture aimed at preventing the unwanted sprouting or regrowth of dormant buds in harvested crops, particularly tubers, bulbs, and rhizomes. Sprouting can occur during storage and transportation due to factors such as exposure to light, fluctuations in temperature, and hormonal changes within the plant tissues. Uncontrolled sprouting not only leads to loss of quality and marketability but also reduces the nutritional value and storage life of horticultural produce. Various methods are employed for sprout inhibition in horticulture, each tailored to the specific requirements of different crops and production systems. Chemical methods involve the application of sprout inhibitors, such as synthetic growth regulators or natural plant extracts, which interfere with the physiological processes responsible for sprouting. Commonly used sprout inhibitors include chlorpropham (CIPC), maleic hydrazide, and ethylene inhibitors, which inhibit cell division and elongation, thereby preventing the emergence of new shoots.

Physical methods, such as temperature management and modified atmosphere storage, are also effective for sprout inhibition by creating unfavorable conditions for sprout development. Maintaining low temperatures and high humidity levels during storage can slow down metabolic processes and suppress sprouting in harvested crops. Additionally, controlling oxygen and carbon dioxide levels within storage facilities can inhibit sprout growth by altering the respiratory activity of plant tissues.

Biological methods, including the use of microbial antagonists and natural predators, offer eco-friendly alternatives for sprout inhibition by targeting specific pathogens or pests responsible for sprouting. Biocontrol agents such as Bacillus subtilis and Trichoderma spp. have been shown to suppress sprout growth by competing for nutrients and space or producing antimicrobial compounds that inhibit fungal growth.

Integrated sprout inhibition approaches that combine multiple control methods are increasingly being adopted in horticulture to enhance efficacy, minimize environmental impacts, and reduce reliance on chemical inputs. By implementing effective sprout inhibition strategies, horticultural stakeholders can extend the storage life, maintain the quality, and ensure the marketability of harvested crops, thereby optimizing post-harvest management practices and promoting sustainable agricultural systems. Tuber and bulb crops, such as onions and potatoes, undergo a period of dormancy after reaching maturity. Sprouting occurs at the conclusion of this inactive or resting phase. Sprouting is the resumption of growth. Sprouting results in significant substrate loss through respiratory use. Maleic hydrazide (MH-40), 3-Chloroisopropyl-N-Phenyl Carbamate (CIPC), Methyl naphthalene acetic acid (MENA), and 2,3,4,6-tetranitrobenzene (TCNB) are frequently employed as substances that prevent the growth of sprouts. Many countries largely recognize gamma irradiation at a dosage of 0.02-0.15 KGY as an effective method to prevent sprouting in onions and potatoes, while not causing any negative effects on other quality characteristics.

Packaging

Packaging plays a pivotal role in the post-harvest management of horticultural produce, serving as a critical link between producers and consumers by ensuring the safe and efficient transportation, storage, and presentation of harvested crops. Effective packaging not only protects horticultural produce from physical damage, moisture loss, and microbial contamination but also helps maintain product quality, freshness, and marketability throughout the supply chain. The selection of appropriate packaging materials and designs is essential in optimizing post-harvest management practices for horticultural produce. Packaging materials vary depending on factors such as the type of crop, transportation requirements, and desired shelf life. Common packaging materials include cardboard boxes, plastic crates, mesh bags, and flexible films, each offering unique advantages in terms of strength, durability, breathability, and barrier properties.

Proper packaging design considers factors such as ventilation, moisture control, and product protection to ensure optimal storage conditions and minimize post-harvest losses. Ventilation holes or perforations in packaging materials facilitate airflow, reducing the buildup of respiratory gases and preventing the accumulation of moisture, which can promote fungal growth and decay. Additionally, moisture-resistant packaging materials help maintain the optimal humidity levels required to prevent wilting, dehydration, and physiological disorders in horticultural produce. Furthermore, packaging plays a crucial role in branding, marketing, and consumer communication, influencing purchasing decisions and product perceptions. Eye-catching designs, informative labelling, and sustainable packaging solutions can enhance the visibility, appeal, and perceived value of horticultural produce, thereby facilitating market access and driving consumer demand.

Advancements in packaging technologies, such as modified atmosphere packaging (MAP), active packaging, and smart packaging systems, offer innovative solutions for extending the shelf life and improving the quality of horticultural produce. MAP involves modifying the composition of the atmosphere within the packaging to slow down respiratory processes and inhibit microbial growth, while active packaging incorporates antimicrobial agents, oxygen scavengers, or moisture absorbers to enhance product safety and freshness. Smart packaging systems utilize sensors and indicators to monitor environmental conditions and provide real-time feedback on product quality and shelf life, enabling timely interventions and informed decision-making throughout the supply chain.

Packaging is an essential and indispensable aspect of managing extremely perishable items. The primary function of packaging is to consolidate the produce into practical pieces for easy handling and protect the produce during distribution, storage, and marketing. The selection of packaging materials is based on the specific characteristics of the plant. It enhances the shelf life of perishable goods and increases their appeal. An effective packaging system safeguards products from any physical, physiological, and pathological degradation during storage, transportation, and marketing. The packaging materials should offer cushioning for fresh produce, which can be achieved using various types such as bamboo baskets, plastic or jute sacks, wooden crates, and corrugated fibreboard (CFB) cartons. Vegetables are primarily packaged using bamboo baskets, gunny bags, and plastic crates. Polyethylene film bags have been shown to be highly effective in decreasing wastage when used to wrap horticultural crops such as capsicum, broccoli, assam lemon, and tomatoes for transportation. The utilization of upright cone baskets in conjunction with dry grass as a packaging material between the layers of fruits can effectively decrease losses in first grade tomatoes from 15 per cent to 3 per cent. The storage life of tomatoes can be extended by packing them in sealed unventilated polyethylene, which creates a modified atmosphere. Printed plastic bags are employed to minimize the amount of light that reaches potato tubers. Onions and potatoes are stored in plastic oven vented bags with a capacity of 25 or 50 kg. The implementation of palletization and containerization will greatly contribute to the establishment of a strong foundation for both domestic and international trade (Nath, 2013). Nath *et al.* (2011) demonstrated the prolongation of the storage period for broccoli, capsicum, and tomatoes by employing various packaging materials and post-harvest treatments.

Packaging plays a multifaceted role in the post-harvest management of horticultural produce, encompassing protection, preservation, branding, and communication. By leveraging appropriate packaging materials and technologies, horticultural stakeholders can optimize post-harvest handling practices, minimize losses, and ensure the delivery of high-quality, safe, and nutritious produce to consumers, thereby supporting sustainable agricultural systems and food security.

Transport

In the intricate dance between farm and table, the post-harvest handling of horticultural produce emerges as a critical juncture, where the delicate balance between preservation and distribution defines the journey from field to fork. This vital stage encompasses a multifaceted array of processes, with transportation standing as a cornerstone. Transport plays an indispensable role in the post-harvest continuum, facilitating the seamless movement of fresh produce from farms to markets, ensuring its availability to consumers across vast geographical expanses. Whether traversing bustling urban centers or rural hinterlands, the efficiency and efficacy of transportation systems profoundly impact the quality, shelf-life, and marketability of horticultural commodities. From the moment crops are harvested to their arrival at distribution centers, the transport infrastructure must navigate myriad challenges, including temperature control, packaging considerations, and logistical intricacies, all while striving to minimize losses and maintain product integrity. Thus, understanding the nuances of transportation in post-harvest handling is paramount, not only for ensuring food security and sustainability but also for fostering economic prosperity within the global agricultural landscape.

Transportation plays a crucial role in the post-harvest process, including processing, storage, and distribution. Horticultural produce is transported from the field to distribution markets via many modes of transportation, including rail, truck, airplane, and ship. Horticultural crops are frequently transported in open trucks in significant amounts. The CFTRI, Mysore has created conical bamboo baskets specifically constructed for stacking and allowing air circulation. These baskets are intended for transporting food by train. Significant losses occur as a result of mishandling, negligent loading and unloading, and the use of inappropriate containers. Transporting product during the cooler nighttime hours using ventilated, insulated evaporative cooled or refrigerated vehicles guarantees the preservation of its quality. Pallets are commonly employed in numerous developed nations for the exchange of horticultural produce. It is crucial to incorporate mechanical loading and unloading, especially by utilizing fork lift trucks. Refrigerated containers, sometimes referred to as reefer containers, are manufactured in developed nations. Encouraging the utilization of containers employing evaporative cooling systems should be promoted in India.

Pre-harvest Sprays for Minimizing Post-harvest Losses in Fruits and Vegetables

Pre-harvest sprays of chemicals have been applied to reduce post-harvest losses in different fruits and vegetables. The quest for minimizing post-harvest losses in fruits and vegetables has spurred innovative approaches, among which pre-harvest chemical sprays have emerged as a significant strategy. These pre-emptive treatments represent a proactive measure aimed at mitigating the myriad factors that contribute to post-harvest deterioration, including microbial spoilage, physiological disorders, and physical damage. By intervening at the pre-harvest stage, these chemical applications seek to fortify the resilience of crops against adverse conditions encountered during harvesting, handling, and transportation. Such interventions often involve the judicious application of pesticides, growth regulators, or other bioactive substances, tailored to the specific needs and vulnerabilities of different crops. However, the utilization of pre-harvest sprays is not without its complexities and considerations, including regulatory frameworks, environmental impact, and consumer safety concerns. Thus, while offering promise in reducing post-harvest losses, the implementation of pre-harvest chemical treatments necessitates a nuanced understanding of their efficacy, sustainability, and implications across the entire horticultural supply chain.

The compound thiophenate methyl, when present at a concentration of 0.05 per cent, was found to be extremely effective in reducing post-harvest losses in Dashehari mangoes. Research has shown that applying a pre-harvest spray of 10 to 15 parts per million (ppm) of giberellic acid (GA3) effectively controls the maturity and delays the ripening of mangoes. This treatment allows for longer storage of the mangoes on the tree. To effectively control post-harvest losses in Nagpur mandarin, it was found that applying three doses of Benomyl, Topsin-M, or Carbendazin

(0.05 per cent) at 15-day intervals before harvest was successful. The application of a 0.6 per cent solution of Calcium chloride (CaCl2) to the grapes, 10 to 12 days before to harvesting, improved their post-harvest lifespan. Applying bavistin (0.5 per cent) as a spray before harvesting led to an additional mango yield of 2,720 kg per hectare. It was shown that treating gulabi grapes with a solution of GA3 at a dosage of 50 ppm resulted in the growth of seedless traits. The application of a 2 per cent urea solution onto banana bunches led to a 2-5 per cent augmentation in the weight of the bunches. Prior to harvesting, decreasing the airflow in greenhouses and submitting tomatoes to heat treatment enhanced their soluble solids content, fruit skin color, and decreased the incidence of chilling injury. Uniform flowering and fruiting in densely planted pineapple crops were successfully obtained by using a solution containing 25 per cent ethrel, 2 per cent urea, and 0.04 per cent sodium carbonate (50 ml). Growing pole beans in a greenhouse led to a doubling of the yield and improved the quality of the pods. In order to minimize the amount of fruits and vegetables lost after harvesting, various compounds such as calcium chloride, calcium nitrate, gibberellic acid-3, 6-BAP, carbendazim, and benomyl can be utilized. These substances, including growth regulators and fungicides, can be utilized either separately or in various combinations.

Marketing System Plays a Pivotal Role in Influencing Post-harvest Losses in Horticultural Produce

The marketing system plays a pivotal role in influencing post-harvest losses in horticultural produce, serving as the bridge between producers and consumers while shaping the dynamics of distribution, pricing, and quality control. In many instances, inefficient or fragmented marketing systems can exacerbate post-harvest losses by introducing delays, inefficiencies, and quality degradation at various stages of the supply chain. Limited access to markets, inadequate infrastructure, and suboptimal transportation networks can result in prolonged transit times and improper handling, increasing the likelihood of spoilage and waste. Moreover, asymmetries in market information and bargaining power may compel producers to offload their perishable goods at lower prices, further disincentiviz.ing investments in quality management and post-harvest technologies. Conversely, well-functioning marketing systems characterized by robust infrastructure, transparent pricing mechanisms, and effective market linkages can help mitigate post-harvest losses by facilitating timely delivery, ensuring fair returns for producers, and incentiviz. ing investments in post-harvest handling practices. By optimizing market access, improving logistical efficiency, and enhancing market transparency, stakeholders can work collaboratively to minimize post-harvest losses and optimize the utilization of horticultural produce, thereby promoting food security, sustainability, and economic development.

Vegetable market is often suffering from several constraints due to their high perishable nature, season market and bulky nature. Assembling and subsequent marketing of the produce is further blocked due to lack of proper storage facilities and quick transport systems. Very often the products are formed to dispose of their produce at a very nominal price where there arise seasonal gluts due to these bottle necks. Another major defect in vegetable marketing is the involvement of several intermediaries which dominate the trade and get huge profit. Consequently, producer's margin in the consumer price becomes very low. It is therefore essential that organized effort for establishing co-operative system of marketing should be enforced at village and district levels to control activity of intermediaries and to regulate the vegetable marketing smoothly and in a streamlined system. Moreover, close co-ordination among Agricultural Marketing Board, National Horticulture Board and state department of agriculture/Horticulture should be ensured to formulate an action plan for regulating marketing of vegetables in a smooth and streamlined way.

Storage of Horticultural Produce and Management of Post Harvest Losses

The storage of horticultural produce and the management of post-harvest losses stand as integral pillars in the intricate tapestry of agricultural supply chains, where the preservation of quality and nutritional value meets the imperatives of food security and economic sustainability. The handling and storage of fruits, vegetables, and other perishable commodities demand meticulous attention to environmental conditions, handling protocols, and preservation techniques to safeguard against spoilage, extend shelf life, and optimize market value. As global populations burgeon and supply chains span ever-greater distances, the imperative to minimize post-harvest losses looms larger than ever, underscoring the need for innovative storage solutions and best practices. Whether through controlled atmosphere storage, cold chain logistics, or novel preservation technologies, the quest to reduce post-harvest losses represents a multifaceted endeavor with far-reaching implications for global food systems, economic resilience, and environmental stewardship. Thus, understanding the nuances of storage management and post-harvest loss mitigation is paramount, not only for ensuring food security and sustainability but also for fostering resilience and prosperity within the horticultural sector.

Proper storage of vegetable produce is crucial for enhancing shelf life, preventing market oversupply, ensuring year-round supply, and maximizing profits for the farmers. The primary objective of storage is to minimize and regulate transpiration, respiration, and disease transmission while simultaneously sustaining vital life processes at the necessary level. Various techniques for storing vegetable products include:

Refrigerated Storage of Fruits and Vegetables

Refrigerated storage of fruits and vegetables is a vital aspect of preserving their freshness, flavor, and nutritional value. Proper storage helps extend the shelf life of perishable produce, reducing spoilage and food waste. The controlled temperature and humidity conditions in refrigerators slow down the natural ripening and decay processes, keeping fruits and vegetables crisp and flavorful for longer periods. Additionally, refrigeration helps maintain the vitamins and minerals present in these foods, ensuring that they remain nutritious even after storage. Understanding the ideal storage conditions for different types of fruits and vegetables is essential for maximizing their quality and optimizing their use in culinary endeavors.

Highly perishable vegetable produce requires refrigerated vegetable storage since it retards the rate of metabolic change, moisture loss, respiratory heat products and spoilage caused by heat production and spoilage caused by micro-organisms and thereby enhances retaining life of vegetable produce. In this method, ambient air is cooled and then passed over the bulk grains via existing aeration system. Refrigerated aeration has been used for cooling dry grain in subtropical climates when ambient temperatures are too high. The initial investment for refrigerated storage system is comparatively higher, but together with the dehumidified air method, it could provide answers to the practicability of aeration for safe commercial storage in tropical climates (Navarro and Noyes, 2002).

☆ **Controlled/Modified atmosphere:** Controlled or modified atmosphere storage (CA/MA) is a sophisticated technique revolutionizing the post-harvest handling of fruits and vegetables. By altering the composition of the air surrounding perishable produce, this method extends shelf life and preserves quality by regulating oxygen and carbon dioxide levels while managing temperature and humidity. CA/MA storage slows down ripening processes, inhibits microbial growth, and reduces physiological degradation, thereby minimizing spoilage and retaining nutritional value. This technology enables growers, distributors, and retailers to optimize storage conditions for different types of produce, ensuring freshness and flavor are maintained for extended periods. With its ability to prolong the availability of seasonal fruits and vegetables and minimize food waste, controlled or modified atmosphere storage plays a pivotal role in enhancing food security, sustainability, and economic efficiency in the agricultural industry. The main purpose of controlled atmosphere (CA) or modified atmosphere is to adjust the atmosphere composition of gases surrounding the commodity by removal or addition of gases. Thus, resulting in an atmospheric composition different from that of normal air. Modified atmosphere does not differ in principle from the controlled atmosphere storage except that the control of gas concentric less precise.

☆ **Hyobaric storage:** Hyobaric storage is a storage method that involves storing product in a partial vacuum, comparable to controlled atmosphere storage. A vacuum is generated by a vacuum pump to achieve a specific required low pressure. The ripening and senescence process can be significantly diminished by reducing the rate of respiration and eliminating ethylene. High-barrier (hyobaric) storage of horticultural produce is a sophisticated method designed to extend the post-harvest life of fruits and vegetables by creating an environment with controlled levels of oxygen and carbon dioxide. This technique involves storing produce in specially designed chambers or containers where the atmosphere can be manipulated to slow down ripening, inhibit microbial growth, and maintain quality. By regulating gas concentrations, temperature, and humidity, hyobaric storage minimizes physiological changes in horticultural products, preserving their freshness, flavor, and nutritional content for prolonged periods. This innovative approach not only reduces food spoilage and wastage but also allows for the distribution of seasonal produce beyond their natural availability, contributing to a more sustainable and efficient food supply chain. Understanding the principles and practices of hyobaric storage is crucial for enhancing food security, promoting economic viability for growers, and meeting the growing demand for fresh, high-quality fruits and vegetables year-round.

✩ **Zero-energy cool chamber:** The development of zero-energy cool chambers represents a significant breakthrough in the management of horticultural produce, offering a sustainable and cost-effective solution for post-harvest preservation. These chambers utilize passive cooling techniques, such as evaporative cooling or thermal mass, to maintain optimal storage conditions without relying on external power sources. By harnessing natural principles of heat exchange and insulation, zero-energy cool chambers effectively regulate temperature and humidity levels, extending the shelf life of fruits and vegetables while minimizing energy consumption and environmental impact. This innovation holds immense promise for small-scale farmers and rural communities, providing access to affordable storage solutions that enhance food security, reduce food loss, and support livelihoods. In tropical regions such as India, there is a significant degradation in the quality of harvested produce due to the absence of on-farm storage facilities. To address this issue, low-cost, environmentally friendly zero-energy cool chambers have been created by IARI in New Delhi. These chambers utilize the principle of evaporative cooling by employing locally accessible materials such as bricks, sand, and bamboo. The temperatures in these rooms are lower than the ambient atmosphere. These chambers are designed for on-site, temporary storage of items at the farmer's field. On-farm storage is necessary to minimize losses in extremely perishable fresh horticulture crops. In response to this issue, cool chambers that are cost-effective, energy-efficient, environmentally friendly, and constructed using locally sourced materials have been developed. These chambers operate based on the principles of evaporative cooling. These chambers, depicted in Figure 14.1, have the capability to consistently maintain temperatures that are 10-15°C lower than the outside environment. Additionally, they can maintain a relative humidity of 90 per cent depending on the time of year. Fruits and vegetables are housed in plastic crates inside the chamber. According to a report (Anon. 2006), the storage time of fruits and vegetables in a cold chamber was found to be extended from 3 days at room temperature to 90 days.

Figure 14.1. Environmentally Friendly Cool Chambers.

☆ **Zero-energy cool chamber:** In tropical regions such as India, there is a significant degradation in the quality of harvested product due to the absence of on-farm storage facilities. To address this issue, IARI New Delhi has created low-cost, environmentally friendly zero-energy cool chambers. These chambers operate based on the idea of evaporative cooling, utilizing readily accessible materials such as bricks, sand, and bamboo. The temperatures in these rooms are lower than the ambient atmosphere. These chambers are designed for on-site, temporary storage of items at the farmer's field. On-farm storage is necessary to minimize losses in extremely perishable fresh horticulture crops. In response to this issue, cool chambers that are cost-effective, energy-efficient, and environmentally friendly have been developed. These chambers are constructed using materials that are readily available locally and operate based on the principles of evaporative cooling. The chambers depicted in Figure 1 have the capability to sustain temperatures that are 10-15°C lower than the surrounding environment. Additionally, they can maintain a relative humidity of 90 per cent depending on the time of year. Fruits and vegetables are housed in plastic crates inside the chamber. According to a report (Anon. 2006), the storage time for fruits and vegetables in the cool chamber increased from 3 days at room temperature to 90 days.

The duration of storage is determined by multiple factors. The factors that influence the quality of these products are variety, stage of maturation, pace of chilling, storage temperature, relative humidity, rate of deposition of CO_2, prepacking, and air-distribution systems. The optimal storage temperature and relative humidity requirements for various vegetables are specified in Table 14.3.

Table 14.3. Optimal Storage Conditions for Key Vegetables in Terms of Temperature and Relative Humidity

Crops	Temperature (°C)	RH (per cent)
Tomato (ripe)	8.5-10.0	85-95
Bottle gourd	8.2-10.5	90-95
Brinjal	8.5-10.0	85-95
Cucumber	8.0-10.5	90-95
Tomato (mature green)	10.0-12.0	80-90
Pumpkin, ginger, sweet potato	11.0-13.0	80-90
French bean, okra	7.0-9.0	85-95
Asparagus, lettuce	3.0-4.0	85-95
Potatoes, tamarillo, lima bean, cowpeas	4.0-5.5	80-90

Post-harvest loss is more severe in comparison to loss during production. Reducing post-harvest losses greatly enhances the availability of vegetables without the need for more land or resources. While it is not possible to totally eliminate losses, they can be minimized by the implementation of contemporary cultural practices, harvesting, handling, marketing, and processing procedures.

Factors Affecting Post-harvest Life of Flower Crops

The post-harvest life of flower crops is influenced by a multitude of factors, spanning from pre-harvest conditions to post-harvest handling practices. Understanding these factors is crucial for maximizing the quality and longevity of cut flowers, which play significant roles in various sectors including floral retail, event decoration, and export industries. Pre-harvest factors such as cultivar selection, growing conditions, and crop management practices can profoundly impact flower quality and shelf life. Post-harvest factors such as temperature management, hydration, and sanitation also play critical roles in determining the longevity of cut flowers. Additionally, the physiological characteristics of different flower species and varieties further complicate post-harvest management, requiring tailored approaches for optimal preservation. By comprehensively addressing these factors, stakeholders in the flower industry can implement strategies to prolong post-harvest life, reduce waste, and ensure that consumers receive flowers of exceptional quality and freshness.

1. Preharvest factors
2. Harvest factors
3. Post-harvest factors

Preharvest Factors

Impacts floral quality in 30-70 per cent of cases. The factors that influence the growth and development of organisms include genetics, light, temperature, soil composition, nutrient availability, relative humidity, seasonal variations, CO2 levels, chemical treatments, irrigation practices, and the presence of pests and diseases.

Harvest Factors

Harvest stage, harvest timing, and harvest method. The harvesting phase of significant flower crops is displayed in Table 14.4.

Table 14.4. Harvesting Phase of Significant Flower Crops

Crops	Commercial Stage of Harvesting
Alstroemeria hybrid	When 1 or 2 florets open in a spike and majority show colour
Anthurium	When on thirds to two third of flowers open on the spadix
Chrysanthemum morifolium	Standard: when outer petal fully expanded. Spray: fully open flower before pollen shadingPompon and decorative: centre of oldest flower fully openAnemone: fully open but before central disc floret begins to elongate
Dianthus caryophyllus	Standard: When flowers are half open at paint brush stageAt least two flowers fully open
Gerbera jamesonii	When outer two row of disc floret show pollen
Gladiolus hybrid	When 2 to 4 lower florets of the spike show colour
Heliconia	When fully open
Lilium hybrid	When he buds are well matured and show colour
Rosa hybrids	Well developed and mature flower buds with one or two petals unfurling from the tip
Tulipa hybrids	Flowers not open but well developed and coloured on the upper half, and rest still green on a long stem
Zantedeschia	The spathe begins to turn downward when the flowers are fully open
Cymbidium hybrids orchid	When almost all buds on the spike open
Tuberose	Single type varieties: when buds are fully open

Post-harvest Factors

Factors such as temperature, light, humidity, water quality, ethylene, sensitivity, preservatives, ventilation, packing, illnesses, and pests can all have an impact on the overall condition and quality of a product.

Storage of Cut Flowers

The storage of cut flowers is a delicate process that involves preserving the freshness, beauty, and longevity of blooms from the moment they are harvested until they reach their final destination. Proper storage techniques are essential for maintaining the quality and market value of flowers, whether they are destined for retail sale, event decoration, or export markets. Factors such as temperature, humidity, hydration, and sanitation play crucial roles in determining the post-harvest lifespan of cut flowers. From refrigeration to specialized packaging methods, various storage approaches are employed to slow down physiological processes, inhibit microbial growth, and minimize physical damage. Understanding the intricacies of cut flower storage is paramount for florists, growers, and distributors alike, ensuring that consumers receive vibrant, long-lasting blooms that bring joy and beauty into their lives. Applying low temperature treatment during storage or transportation effectively decreases the overall metabolic activity in the tissues, resulting in a slowdown of respiration, transpiration, and ethylene action. Additionally, it inhibits the growth of bacteria and fungi, preventing their multiplication and subsequent damage.

- ☆ **Cold storage:** This is a conventional technique for preserving freshly cut flowers. Typically, flowers that thrive in moderate climates such as roses and carnations are stored at temperatures ranging from 0 to 10 degrees Celsius. On the other hand, flowers that prefer subtropical climates like gladiolus, sterlitzia, jasmines, proteas, and gloriosa are stored at temperatures between 4 and 70 degrees Celsius. Lastly, flowers that thrive in tropical climates like as anthurium, cattleya, vandal, and euphorbia are stored at temperatures ranging from 7 to 15 degrees Celsius.

- ☆ **Wet storage:** In this technique, flowers are immersed in water or a preservation solution, with their stems submerged, for a brief period. This is appropriate for flower species that have been cut at a commercial stage and are intended to be sold straight within a timeframe of 1 to 2 days. During wet storage, flowers are maintained at a temperature range of 3 - 40C, which is slightly higher than the temperature used for dry storage. This approach provides protection for flowers against fungal infection. To prevent moisture and eventual decomposition, the lowest leaves are cut off the stem. The wet storage period and optimal temperature for some cut flowers are displayed in Table 14.5.

- ☆ **Dry storage:** This is utilized for the extended preservation of flowers. In this technique, newly picked flowers are gathered in the morning, sorted according to quality, and then carefully packaged in plastic bags or boxes to prevent any moisture loss. This approach conserves storage space. Flowers should undergo fungal treatment before being stored. The flowers can be treated with floral preservatives that include sugar, antimicrobial agents, and anti-ethylene chemicals. Table 6 displays the optimal temperature for storing cut flowers and the maximum duration for storage.

  Wet storage period and optimum temperature of some cut flowers and recommended dry storage temperature and maximal period of storage of cut flowers are given in Tables 14.5 and 14.6 respectively.

Table 14.5. Wet Storage Period and Optimum Temperature of some Cut Flowers

Cut Flower	Temperature (°C)	Period
Antirrhinum	4	4 weeks
Carnation	4	4weeks
Chrysanthemum	1	3 weeks
Gladiolus	0.5 -1.54- 6	2-4 days
Gerbera	4	3- 4 weeks
Lily	0-1	6weeks
Rose	42	10 days
Tulip	- 0.5 - 0	2-3 days

**Table 14.6. Recommended Dry Storage Temperature and
Maximal Period of Storage of Cut Flowers**

Cut Flower	Temperature (°C)	Period
Anthurium	13	4 weeks
Carnation	0-1	16-24 weeks
Cattleya	7-10	2 weeks
Dendrobium	5 -7	2 weeks
Gladiolus	4	4 weeks
Lily	1	6 weeks
Rose	0.5 -321	2 weeks4 weeks3 weeks
Tulip	0 - 1	8 weeks

☆ **Controlled atmospheric storage (CA):** In this storage method, fresh flowers are placed in airtight refrigerated chambers that are equipped with chilling equipment. The chambers are maintained at elevated levels of carbon dioxide (CO_2) and reduced levels of oxygen (O_2). This is done to decrease the rate of respiration and minimize the formation and effects of ethylene. Table 14.7 displays the controlled atmosphere (CA) storage conditions for several cut flowers (Table 14.7).

Table 14.7. CA Storage of some Cut Flowers

Flowers	CO_2 (per cent)	O_2 (per cent)	Storage Temperature (°C)	Storage Period (Storage)
Carnation	5	1.3	0-1	30
Freesia	10	21	1-2	20-22
Gladiolus	5	1-3	1.5	21
Lily	10-20	21	1.0	21

Flowers	CO_2 (per cent)	O_2 (per cent)	Storage Temperature (°C)	Storage Period (Storage)
Rose	5-10	1-3	0	20-30
Tulip	5	21	1	10

☆ **Low Pressure Storage:** Plant materials are preserved in a vacuum environment, at a low temperature, and with cooled humid air. During Low Pressure Stress (LPS), plants or their organs release gaseous compounds, such as CO_2 and ethylene, at an accelerated pace through stomata and intercellular gaps, as opposed to normal pressure conditions. The positive effect of LPS is attributed to the reduction in ethylene production at low O_2 levels, as well as the inhibition of other volatile compounds. Table 14.8 presents the specific storage conditions for particular cut flowers and herbaceous cuttings in a low-pressure environment.

Table 14.8. Low Pressure Storage of some Cut Flowers and Herbaceous Cuttings

Cut Flowers	Storage Period (Days)	
	Cold Storage	Low Pressure Storage
Carnation	10	91
Chrysanthemum	7-14	2
Rose	7-14	56
Unrooted cuttings		
Chrysanthemum	10-28	42-94
Carnation	10-20	300
Geranium	5-10	21-28
Poinsettia	5-103	21-283
Rooted cuttings		
Chrysanthemum	7-14	90
Geranium	14	28
Poinsettia	7	14

☆ **Modified Atmospheric Storage (MA):** Modified Atmospheric Storage (MA) is a pioneering method in the realm of horticultural produce preservation, revolutionizing the post-harvest landscape by creating tailored environments to extend the shelf life and maintain the quality of fruits and vegetables. By adjusting the composition of gases surrounding the produce, particularly oxygen and carbon dioxide levels, MA storage retards ripening, slows down enzymatic activity, and inhibits microbial growth, thereby preserving freshness and nutritional content. This innovative approach not only minimizes food waste but also facilitates the year-round availability of seasonal produce, meeting consumer demands while optimizing supply chain efficiency. Understanding the principles and implementation of MA storage is instrumental for growers, distributors, and retailers in enhancing food security, sustainability, and economic viability in the agricultural sector. MA storage is less precise form of CA storage. The dry storage of flowers in sealed bags leads to reduction in O_2 and increase in CO_2 levels due to respiration of the tissue.

Processing and Value Addition in Vegetables and Flowers

The processing sector of horticultural crops, which includes vegetables and decorative crops, plays a crucial role in managing surpluses and reducing waste in the horticulture industry. Processing can provide growers with an extra source of income and contribute to price stabilization by generating economic rewards. The most reliable measure of the economic impact of food processing on the food system is the value added. The measure of the industry's contribution to GDP is determined by the value added. To mitigate market oversupply and reduce post-harvest losses, vegetables can be processed, preserved, and marketed during the off-season, in addition to the peak season. A wide range of value-added products can be made from different vegetables, including ready-to-cook vegetables, tomato soup, jam, candy, canned peas, tomato sauce and ketchup, puree and paste, frozen and dehydrated capsicum, cabbage, and French beans, as well as ginger and turmeric products such as oil, oleoresin, powder, and pickles (Nath *et al.,* 2016).

Low-Cost Processing Technologies for Small and Marginal Farmers

Empowering small and marginal farmers with low-cost processing technologies is not just an economic imperative but a transformative step towards enhancing agricultural productivity, improving livelihoods, and fostering rural development. These farmers, often operating on limited resources and facing numerous constraints, require access to affordable and efficient processing solutions to add value to their produce, reduce post-harvest losses, and tap into higher-value markets. Low-cost processing technologies encompass a range of innovative methods and tools designed to streamline operations, increase efficiency, and improve the quality and marketability of agricultural products. From simple techniques like solar drying and traditional preservation methods to affordable machinery and equipment for sorting, grading, and packaging, these technologies empower small-scale farmers to compete in today's dynamic agricultural landscape. By investing in accessible processing solutions tailored to their needs, policymakers, development agencies, and agricultural stakeholders can unlock the untapped potential of small and marginal farmers, driving inclusive growth, food security, and sustainable rural development.

Processing of Chow-chow (*Sechium edule* SW.) into Tuity Fruity

The chow-chow, scientifically known as Sechium edule SW., is a popular indigenous vegetable that is extensively cultivated in the mountainous regions of India. This crop may be cultivated with few inputs, resulting in an average yield of 120 fruits per plant and 30-40 tons per hectare. During the May to September season, it is abundantly accessible on the market at a fairly affordable price. Currently, there are no value-added products derived from this vegetable, resulting in a lack of demand for it in the food processing sectors. Consequently, the low price of this crop serves as a deterrent for farmers to engage in large-scale cultivation.

A technique for producing a homemade chow-chow fruit preserve can be achieved by utilizing mature chow-chow veggies (Nath *et al.,* 2014). For this process, the veggies that have been peeled should be sliced into 10 x 10 x 5 mm pieces and then submerged in boiling water for 5-10 minutes. Immerse blanched vegetables in a sugar syrup solution with a 40 per cent concentration for 1 hour. Afterwards, it is necessary to submerge them in a sugar syrup solution with a concentration of 70 per cent for an extra duration of 2 hours. Throughout this procedure, the application of heat should be carried out in a gradual manner. In addition, the solution should consist of 15-20 per cent juice derived from the Sohoing plant (Prunus nepalensis) and 10-15 per cent juice obtained from fresh ginger (Gingiber officinale). These juices function as natural agents for coloring, acidifying, and flavoring. After the required duration of boiling has passed, the slices should be drained and subsequently subjected to a tray drier for 30 minutes at a temperature of 50°C. The resulting chow-chow tuity fruity products may have a total soluble solids (TSS) percentage ranging from 70 per cent to 73 per cent. The product should be packaged using suitable materials to prolong its shelf life. This can be employed for confectionery, bakery, and pan masala applications. Below is a flow chart that demonstrates the method of preparing chow-chow tuity fruits.

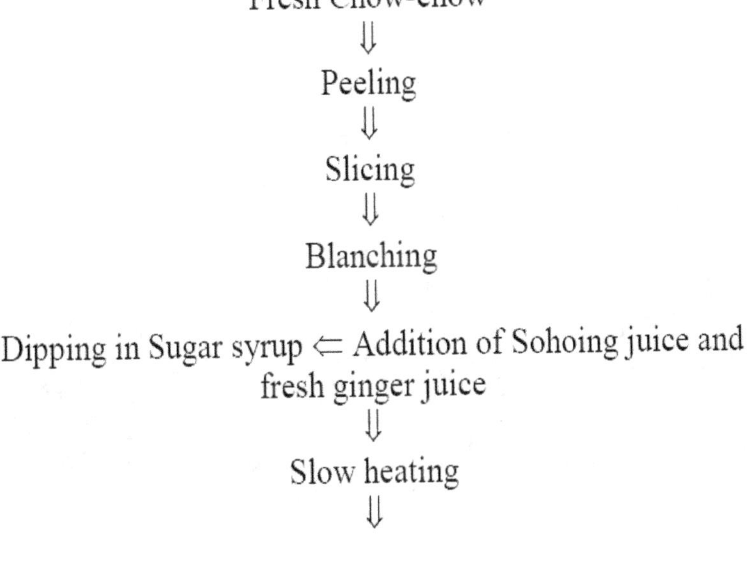

Blanching

⇓

Dipping in Sugar syrup ⇐ Addition of Sohoing juice and
fresh ginger juice

⇓

Slow heating

⇓

Production of Beet Root Tuity Fruity, Nectar and Jam

Red beetroot, scientifically known as Beta vulgaris, is an excellent natural food colorant due to its substantial concentration of red-colored betalain pigments. There is a growing fascination with the utilization of natural food colors, which offer health advantages such as antioxidant properties. The presence of phenolic compounds in beetroot is responsible for its antioxidant capacity, which contributes to the promotion of human health and the prevention of degenerative diseases and cancer, providing nutraceutical benefits. Red beetroot, recently grown, were obtained at the local vegetable market, Meerut, UP, in February 2015. They were then washed with clean water. The beetroots were peeled and sliced using a stainless steel knife. The RSM design takes into account various slice thicknesses ranging from 2 to 12mm and blanching times ranging from 0 to 15 minutes. The blanched slices were immersed in a sugar syrup with a concentration of 60 per cent and subjected to gradual heating for a duration of 2 hours. Various quality measures, such as organoleptic scores, antioxidant content, betalain, ascorbic acid, and phenol content, were assessed. The beetroot treatment that used slices with a thickness of 8 mm and a blanching period of 4 minutes achieved the highest scores in overall acceptability (8.1), taste (8.3), and texture (7.8). The beetroot nectar was made consistent by using 25 per cent beetroot juice, 0.3 per cent acidity, and 22 per cent TSS in the finished goods, ensuring they had a satisfactory color and taste. Beetroot jam was standardized by combining boiled beetroot juice with water in a 1:2 ratio, along with 1.2 per cent pectin powder, 0.4 per cent citric acid, and a final TSS (Total Soluble Solids) of 69 per cent. Beet root tuity fruity, Beet root Nectar and Beet root jam are presented in Figures 14.2–14.4 respectively.

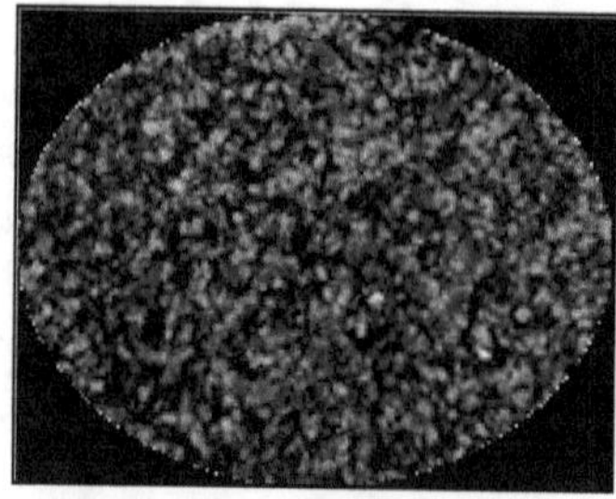

**Figure 14.2. Beet Root
Tuity Fruity.**

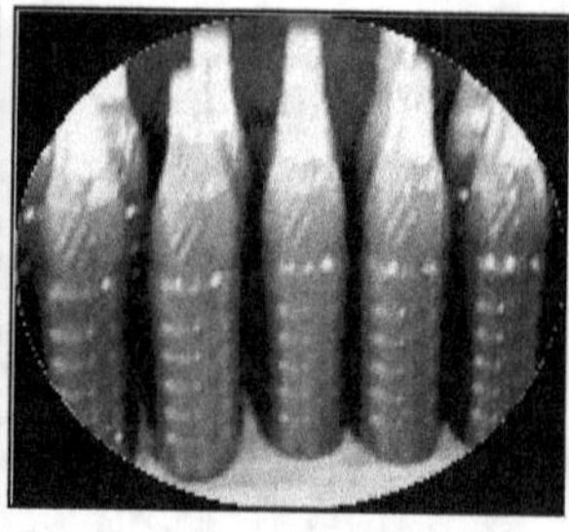

**Figure 14.3. Beet Root
Nectar.**

**Figure 14.4. Beet Root
Jam.**

Instant Ginger Candy

Nath *et al.* (2013) enhanced the qualitative characteristics of instant ginger candy by varying the thickness of the slices and the duration of blanching. For making candy, it is recommended to use ginger rhizomes of the cultivars Nadia or Baroda that are six months old and have a consistent size. It is necessary to cleanse materials meticulously using fresh water in order to eliminate dirt and other undesired particles from the surface, as well as to minimize the presence of microorganisms that might cause contamination. After washing, the rhizomes should be dried at room temperature for a few hours. Peel the dried fresh rhizomes by hand and use a stainless-steel knife to make slices of varying thickness (ranging from 5.0 to 25.0 mm). The slices should be immersed in boiling water for 25-30 minutes, and then soaked in sugar solutions with a concentration of 40°B and 75°B, respectively. These sugar solutions should also contain 2.0 per cent citric acid. The soaking times for the slices in the sugar solutions are 1 hour and 2 hours, respectively. The temperature for this process should be maintained at 95°C. Once the retention period has reached the predetermined level, the materials should be removed from the syrup and placed in a laboratory tray drier at a temperature of 60°C for a duration of 1 hour. Prior to being placed in air tight containers for further analysis, it is necessary to chill dried materials at room temperature. The flowchart illustrating the process of preparing instant ginger candy is displayed below:

Freshly harvested uniform size ginger rhizomes

⇩

Washing with clean water

⇩

Peeling

⇩

Cutting into round shaped slices

⇩

Blanching in boiling water
(25-30 minutes with 2.0% citric acid)

⇩

Dipping in 40°Brix sugar syrup with 2.0% citric acid
(for 1 hr at 95°C)

⇩

Dipping in 75°Brix sugar syrup with 2.0% citric acid
(for 2 hrs at 95°C)

⇩

Draining and drying at 60 °C for 1 hour

⇩

Cooling and packing

Preparation of Ginger Slice in Brine

This is a lightly processed fresh ginger product that can be stored for over six months under normal conditions. This product is often made from fresh and tender ginger rhizomes. The rhizomes should be peeled and sliced into 5-10 mm thick pieces. These slices should then be blanched in boiling water at 100°C for 5-10 minutes. After blanching, the slices can be filled into bottles or cans and covered with a brine solution containing 9.0 per cent salt and 2.0 per cent citric acid. In order to enhance the longevity of the product, it is possible to use preservatives such as 50 ppm of potassium metabisulfite (KMS) and 50 ppm of sodium benzoate in the brine solution. Securely seal the bottles and store them in a cool location.

Recipe for Ginger Slice in Brine

Sl.No.	Raw Materials	Quantity
1.	Sliced ginger (5mm thickness)	1.0 kg
2.	Water	1.0 liter
3	Salt	90.0 g
4.	Citric acid	20.0 g
5.	KMS	0.05 g
6.	Benzoic acid	0.05 g

The flowchart for the preparation of ginger slices in brine is given below:

Fresh Ginger
⇩
Peeling
⇩
Slicing
⇩
Blanching (5-10min at 100°C)
⇩
Filling into bottles/cans
⇩
Covering with brine solution
⇩
Sealing/seaming
⇩
Storage

Preparation of Ginger Paste

This ginger product is made from various cultivars that have a low fiber level, making it suitable to use. The following steps are crucial in the manufacture of ginger paste:

- ☆ Select young and fibrous-free ginger rhizomes that are soft in texture.
- ☆ Remove the outer skin of the rhizomes and ground them into a smooth paste using a mixer grinder.
- ☆ Incorporate 3-5 per cent of common salt into the paste and mix it well. A small quantity of water can be added, if necessary, to achieve the appropriate consistency.
- ☆ Heat the entire mixture for 10-15 minutes while continuously stirring.
- ☆ Lastly, include acetic acid and KMS into the paste and transfer it into a glass jar or plastic container.

Recipe for Ginger Paste

Sl.No.	Raw Materials	Quantity
1.	Ginger pulp	1.0 kg
2.	Salt	30-40.0 g
3.	Acetic acid	15-20.0 ml
4.	KMS	1.0 g
5.	Water (if needed)	100.0 ml/kg

The flowchart for the preparation of ginger paste is given below:

Fresh ginger
⇩
Peeling
⇩
Pulping
⇩
Heating for 10-15 minutes
(Addition of salt & water)
⇩
Addition of Acetic acid & KMS
⇩
Heating
⇩
Filling in bottles

Tomato Ketchup

The product is composed of tomato juice or pulp that has been strained, along with spices, salt, sugar, and vinegar. It may or may not include onion and garlic. The product must have at least 12 per cent tomato solids and 25 per cent total solids. The following steps illustrate the various stages involved in the preparation of tomato ketchup:

Processing Steps

- ☆ Rinse the tomatoes thoroughly with clean water
- ☆ Submerge the tomatoes in boiling water for 20 minutes
- ☆ Peel off the skin and extract the juice/pulp
- ☆ Combine one third of the amount of sugar with the extracted tomato juice and begin heating
- ☆ Place all the necessary ground spices in a muslin cloth (spice bag) and add it to the pulp while continuing to heat
- ☆ Heat the mixture until the volume of the pulp reduces to one third of its original volume
- ☆ Remove the spice bag and discontinue the heating process. Incorporate vinegar, salt, and the remaining sugar into the mixture, ensuring thorough mixing. Reheat the mixture slightly to thoroughly blend the ingredients.
- ☆ Incorporate the precise quantity of sodium benzoate into the product
- ☆ Transfer the product into sterilized bottles and allow it to cool until it reaches a temperature of around 40°C
- ☆ Securely seal the bottles and affix a label indicating "Store in a dark and cool place."

Recipe for Tomato Ketchup

Ingredients	Amount	Ingredients	Amount
Tomato juice	1 kg	Sugar	200-250 gm
Salt	10 gm	Onion	20 gm
Garlic	1 gm	Cinnamon	1.0 gm
Black Pepper	1 gm	Ginger	5.0 gm
Clove	2 no.	Red chilli	5.0 gm
Vinegar	5 per cent	Sodium benzoate	750 ppm (0.75 gm per liter)

Preparation of Low-Cost Tomato Powder

It is composed of fully developed and ripe fruits. This product is commonly utilized for the production of several tomato-based items such as juice, soup, ketchup, puree, chutney, and so on. The following stages outline the process of preparing tomato powder:

Processing Steps

- ☆ Take fully ripe tomatoes
- ☆ Wash with clean water
- ☆ Make slices of 5.0-7.0 mm thickness
- ☆ Place the slices in single layer in the aluminum trays
- ☆ Dry the slices in Cabinet drier at 60°C for 7-8 hours till slices becomes crisp
- ☆ Cool the slices and grind with laboratory grinder to converted into free-flowing powder
- ☆ Pack the powder either in air tight plastic containers or flexible plastic packaging materials
- ☆ Store the product in dark, dry and cool climate for future use

Value Added Products of Commercial Flowers

The commercial flower industry thrives not only on the beauty and fragrance of blooms but also on the diversity of value-added products derived from these natural treasures. From elegant floral arrangements adorning homes and events to essential oils capturing delicate fragrances, the possibilities are as vast as the array of flowers themselves. This flourishing industry not only caters to aesthetic preferences but also meets practical needs, offering everything from therapeutic extracts to culinary delights. Through innovative processing techniques and creative craftsmanship, commercial flowers transcend their ornamental role, becoming integral components in an array of products that enhance our lives in myriad ways. In this exploration, we delve into the captivating world of value-added products derived from commercial flowers, uncovering the ingenuity and artistry that transform petals into an array of exquisite goods with both economic and cultural significance.

- ☆ **Rose water:** Rose water is produced through the process of boiling rose blossoms in water and then condensing the resulting vapor. In India, a rudimentary technique involves the utilization of a copper kettle (known as Deg) and a condenser made of bamboo, which is continuously kept moist by being covered with rope. This process mostly produces Attar and rose water.

- ☆ **Rose oil, ottoo or attar of roses:** The attar is composed of the remaining eleven essential oils, as well as wax extracted from the petals' surface and other small contaminants. It is the most expensive among the natural essential oils used in high-quality fragrance goods worldwide. The otto of rose is not derived directly from the distillate, but rather from the rosater that is collected during the distillation process of rose petals. Rose water, when held in earthen pots or steel vessels during cool nights in an open area, accumulates a fragrant substance resembling butter called otto, which floats on the surface of the water.

☆ **Concrete:** This is a hydrocarbon solvent extract obtained from fresh flowers. The substance contains a high concentration of hydrocarbon-soluble substances and does not contain any chemicals that dissolve in water. The substance is typically a waxy, partially solid, dark-colored compound that is devoid of any initial solvent.

☆ **Absolute:** An alcoholic extract of concrete, highly concentrated and consisting solely of alcohol soluble particles, primarily used in alcoholic perfumes.

☆ **Itra of rose:** Rose oil is obtained through the process of distilling rose blossoms and absorbing the resulting vapors in sandalwood oil. The popular attars that are often sought after include rose, jasmine, heena, kewda, and others.

☆ **Pomade:** It is derived from a technique called effleurage, which is mostly practiced in France. The effleurage technique is mostly employed to extract aromatic substances from delicate flowers that contain small amounts of aromatic compounds and cannot endure other processing methods, such as exposure to heat or steam. Effuerage is a method of extraction that involves using frigid temperatures. The fat is covered with either fresh petals or intact flowers, and the frames are then placed into stacks. The process of manufacturing jasmine pomade can take between 12 and 30 hours, whereas tuberose pomade might take between 24 and 100 hours.

India's diverse agroclimatic conditions, varied soil types, and abundant rainfall provide great opportunities for cultivating a wide range of horticultural crops. These include fruits, vegetables, flowers, plantation crops, tuber and rhizomatous crops, as well as crops with medicinal and aromatic properties. India ranks second globally in terms of fruit and vegetable production, following China. The export of horticultural produce, notably flowers, contributes significantly to the country's earnings. Regrettably, around 25-30 per cent of horticulture produce, 10-25 per cent of vegetables, and 30-40 per cent of flowers are wasted owing to insufficient post-harvest management, resulting in significant financial losses amounting to crores of rupees. To mitigate these post-harvest losses, the shelf life of fresh horticulture food can be extended through before or post-harvest management measures or by processing it into various value-added products. Various factors contribute to post-harvest losses in fruits and vegetables, such as physical, physiological, mechanical, and unsanitary conditions, inadequate storage conditions, absence of refrigerated facilities, and the presence of diseases and pests. During the process of harvesting, handling, storage, and marketing, many wound pathogens can infect the produce, leading to the deterioration of its quality and quantity, finally resulting in economic losses. The degradation of fruits and vegetables after they have been harvested can happen either between the time they flower and reach maturity, or during the process of harvesting and subsequent handling and storage. Several methods have been invented in the past and documented in the literature, but they are not widely used, possibly due to factors such as limited local availability of materials, lack of effectiveness, or high cost. Implementing basic post-harvest management techniques such as correct harvesting, sorting, grading, packaging, pulping, pickling, drying, and dehydration at the farmer's level during the peak season will effectively reduce post-harvest losses.

Glossary

Abscisic acid: Abscisic acid is a plant hormone that induces dormancy in buds and seeds.

Acclimation: Plant acclimatization is the process of adjusting plants to a novel environment, which holds significant relevance for newly introduced plants.

Acidic Loving Plants: Plants that thrive when the pH level is less than 7. Soil amendments such as sulfur may need to be introduced.

Adaptable as a Houseplant: This indicates that the plant has the ability to be cultivated indoors, particularly during the winter season, and maybe throughout the entire year.

Adaptation: Plant traits that enable it to thrive in specific environmental conditions. Regional factors. Certain plants possess the ability to adapt and thrive as houseplants, specifically being able to grow indoors throughout the winter season.

Aeration: Performing mechanical soil aeration to promote optimal air and water circulation.

Aerial application: Fertilizer solutions are applied by airplane in regions where ground application is not feasible, such as steep terrain, forest lands, grasslands, or sugarcane fields.

Aeroponic cultivation: Aeroponic cultivation is a method of cultivating plants where they are exposed to a misted environment and their roots are suspended in the air. The roots are irrigated with water that is rich in nutrients, which supplies the essential elements required for growth. This approach demonstrates exceptional efficiency in terms of water utilization and yields accelerated growth rates.

Alkaline: Plants that necessitate a pH level higher than 7. Soil amendments such as lime may be necessary to add.

Auxin: A crucial phytohormone that regulates plant development.

Bark: The outer layer of a woody trunk primarily composed of non-living material. Provides protection for the cambium.

Basal dressing: Basal application through broadcasting refers to the act of spreading fertilizers before sowing or planting crops and mixing them by cultivating the soil during seed bed preparation.

Biennial: These plants complete their life cycle after two years.

Bio- fertilizers: Biofertilizer is an organic substance that contains a particular bacterium capable of converting nutrients that are not readily available into a form that is accessible through biological processes. Examples of such microorganisms are Rhizobium, Azotobacter, Azospirillum, Blue Green Algae, and PSB.

Blight: A plant disease that kills leaves, flowers and stems.

Bower system: The system is alternatively referred to as 'Pandal' or 'Pergola'. The method is commonly observed in grapes and other cucurbitaceous vegetables such as snake gourd, ribbed gourd, and bitter gourd. In this design, the vines are arranged in a grid-like pattern of wires, often positioned at a height of 2.1 to 2.4 meters above the ground. These wires are upheld by either concrete or stone pillars, or by living support structures such as Commiphera sp. The vine is permitted to grow a solitary stem until it reaches the wire mesh and is typically upheld by bamboo stakes fastened with jute twine. Once the vine reaches the wires, its apex is pruned to encourage the growth of lateral branches.

Bract – Modified leaf that are colorful. Example, dogwood and poinsettias.

Brambles – Plants with cane shoot growth. Example, raspberries and blackberies.

Bridge grafting: Bridge grafting is performed with the aim of restoring a damaged fruit plant. The scions are prepared by making oblique incisions on one side of the top and bottom. The scions are put both above and below the damage of the plant and securely secured.

Broadleaf evergreen: A shrub or tree with green leaves (as opposed to needles) that hold their foliage all year long. Example, Euonymus fortune.

Bud – Embryonic plant tissue.

Budding: Budding is a grafting technique that involves using a single bud together with a piece of bark, with or without wood, as the scion material. This scion material develops into a plant once it successfully joins with the stock. Budding is often performed during the period when the stock plant is experiencing active growth and increased cambial activity.

Bulb – Underground modified leaves that form roots. Examples, onions and tulips.

Cambium – A thin tissue layer that provides cells for plant growth. It is found between the xylem and phloem. If it becomes severely damaged the plant will probably die.

Canker – Open wound stem, branch or trunk injury that becomes infected with fungal or bacterial pathogens.

Catkin – Drooping flower clusters. Examples, willow,birch and oak.

Center Pivot Irrigation: An automated irrigation system that uses rotating sprinklers mounted on wheeled towers to water circular fields.

Chlorophyll – The pigment that gives plants green coloration.

Chlorosis – A yellowing of leaf tissue due to the lack of chlorophyll. Iron or magnesium additives are used to treat symptoms.

Compaction: Soil pressure that prevents water and air flow.

Companion crops: Two noncompetitive crops grown at the same time in the same area.

Compost: Compost refers to the process of breaking down plants and other previously living materials into a soil-like substance. This substance is rich in organic matter, serves as a very effective fertilizer, and has the ability to enhance the quality of nearly any type of soil.

Container Plant Style: Plants utilized in combinations are occasionally categorized as teasers, fillers, and spillers to designate the specific role each plant plays in a combination design. This term refers to the style of using plants that are grown in containers.

Controlled Release Fertilizer: Time Release Fertilizer is another term used to refer to this product. Fertilizer is available in pellet form and is an enhanced iteration of Slow-Release Fertilizer. Fertilizer is released solely depending on soil temperature, rather than microbial activity, and generally exhibits more precision compared to Slow-Release Fertilizer.

Coppicing: This procedure involves the total removal of the main stem of trees such as Eucalyptus and Cinchona, leaving only a stump measuring 30-35cm. The coppiced stump begins to generate several robust shoots after around 6 months. Only 2-3 shoots are preserved per stump while the remaining ones are completely removed. The neglected shoots reach the stage of coppicing in around 10 years, depending on the specific sites and other variables.

Cut Flower: Cut flowers are fresh flower harvested inclusters/spike or in single along with their stem.

Cutting: Production of a new plant from a cutting.

Cutting: It is a way of propagation where any vegetative element of the plant is separated and planted to regenerate the missing parts and develop into a new plant. This method is known as detached propagation. This technique is frequently employed in plants that exhibit high rooting capacity and rapid multiplication, resulting in a cost-effective and expeditious plant propagation.

Cytokinin: An essential plant hormone promoting cell division.

Damping off: Seedling disease causing decay prior to germination.

Dead heading: The removal of old and dying flowers which can promote new growth.

Deciduous: Plants that shed all their leaves at the end of a growing season usually followed by dormancy.

Desiccation: Drying out and dying of plant tissue.

Dicot: Two cotyledons present in the seed.

Dioecious: Where dioecious plants exhibit sexual dimorphism with distinct male and female blooms on individual plants. Pollination necessitates the presence of both a male and female plant. Illustration: ginkgo.

Disbudding: Disbudding is the act of selectively removing undesired flower buds within a cluster in order to promote the growth of the remaining buds into a single, impressive, and high-quality bloom. This procedure is frequently carried out in cut flowers such as carnation, chrysanthemum, dahlia, marigold, and zinnia, among others.

Division: A method of propagation by dividing plant roots.

Drip Irrigation: Hydroponic irrigation is a technique that involves the precise distribution of water to the root system of plants using a system of interconnected tubes, pipelines, and emitters. This method guarantees optimal water utilization.

Drip zone: The area around a tree where water drops from the canopy. This is a useful area for root feeding.

Dwarf: A small plant or plant variety as compared to the parent plant.

Economic Flowers: Economical flowers refer to flower harvests that are cultivated as field crops in specific regions of the State on a commercial scale, with the aim of supplying a significant quantity of flowers to the market.

Evergreen: Leaf retention and functionality is maintained throughout the year.

Exotic: A plant not native to a particular region.

Fertigation: Drip irrigation is the process of delivering water and fertilizers directly to the root zones of plants simultaneously.

Filler: These plants occupy the central space of a container, bridging the gap between the trailing plants and the plants that provide volume, resulting in a visually complete pot.

Floriculture: Floriculture is a branch of ornamental horticulture that focuses on the cultivation and utilization of flowers, potted plants, and annual bedding plants. It involves the utilization of flower items in the florist industry. Poinsettias, carnations, philodendrons, and petunias are frequently encountered plant species in the field of floriculture.

Flower: The seed bearing part of a plant.

Foliar feeding: Applying, usually by spray, fertilizer to plant leaves.

Frass: Termite droppings. A sign that termites are present.

Frost line: The depth at which soil groundwater freezes.

Frost-Free Date: The average date in spring when your area reaches the point where it no longer experiences freezing temperatures and the average date in fall when your area encounters the initial occurrence of freezing temperatures. This date is crucial for determining the optimal time to sow seeds in the spring. Having knowledge of both the dates of spring and fall frosts will assist you in determining the duration of your growing season.

Fruit: The ripened plant ovary which contains seeds.

Fruiting habit: Describes the way in which fruit forms and grows on woody plants.

Fungicide: Pesticides that kill and prevent fungi and their spores.

Gibberellin: An important plant hormone that promotes internode elongation and cell enlargement.

Girdling: The process of removing woody material from the inner layer of the tree called the cambium. Ring-barking is another term used to refer to this phenomenon. Ultimately, it has the potential to be lethal to the plant.

Graft union: The point where grafted plants join together. Also called a bud union.

Grafting: Grafting is the process of uniting plants to create a single cohesive growth structure. For instance, the majority of roses that are sold in the market are grafted onto rootstock.

Greenhouse cultivation: Greenhouses are enclosed structures constructed from glass or plastic that provide a regulated environment for the cultivation of plants. They enable the regulation of temperature, humidity, and light, creating an optimal environment for plant cultivation.

Hair Cut: Pruning a plant involves using a precise tool such as sharp scissors or shears to trim the ends of the branches. This process ensures that an equal quantity is cut from all regions of the plant, similar to how a hairdresser trims your hair. If there are elongated segments extending or protruding, they should be pruned more extensively to achieve a uniformly balanced appearance of the plant. Trimming the plant will tidy its appearance and promote the growth of additional branches.

Headspace: Gap between the soil surface and the container rim. This gap serves to prevent soil erosion and facilitate the flow of water into the container while watering. In the absence of head room, water might readily overflow from the container.

Heavy Feeders: Plants that require a substantial amount of fertilizer to achieve their best performance. Consistent application of fertilizer is essential for maintaining ongoing performance.

Herbaceous: Herbaceous plants lack a persistent woody framework. Herbaceous perennials are plant species that die back to soil level at the end of each season. Some examples of flowers include coneflower, day lilies, and coreopsis.

Herbicide: A chemical, usually in spray form, that kills weeds.

Hi-tech horticulture: Hi-tech horticulture is a cutting-edge technology that is currently popular. It is less reliant on the environment and requires a significant amount of capital investment. However, it has the potential to greatly increase productivity and improve the financial situation of farmers. Hi-tech horticulture has advantages not only for cultivating crops such as fruits, vegetables, and flowers, but also for purposes such as conservation, plant protection, post-harvest management, and adding value to the produce.

Hormone: Plants create vital compounds that control various elements of their life cycle, including growth, development, reproduction, and longevity.

Horticulture: Horticulture is a distinct division within the broader agriculture business. Horticulture is the practice of cultivating a garden. Nevertheless, the phrase has acquired a more expansive connotation. Horticulture encompasses the cultivation and utilization of plants for the purposes of sustenance, convenience, and enhancement of aesthetic appeal.

Hydroponic cultivation: Hydroponic horticulture involves growing plants without soil, instead using a medium that lacks dirt and exposing the roots to a water solution that is rich in nutrients. This technique enables meticulous regulation of nutrient levels, pH, and water supply, leading to increased crop production and accelerated growth.

Hydroponics: Hydroponics is a horticultural technique that cultivates plants without the need for soil.

Integrated Disease Management(IDM): Integrated Disease Management (IDM) refers to the consolidation of various disease control techniques. A combination of physical, chemical, and mechanical approaches is employed to eliminate various diseases in different crops.

Integrated nutrient management(INM): Integrated nutrient management (INM) is a method used to maintain soil fertility and provide optimal plant nutrient supply for maximum crop output. It involves maximizing the advantages from all available sources of plant nutrients in a coordinated manner. Advanced horticulture relies significantly on the prudent utilization of water and fertilizers. Due to restricted land availability and the need to boost production, the soil will become incapable of supplying the necessary nutrients for crop multiplication. It is imperative to utilize fertilizers in optimal ratios.

Integrated Pest Management (IPM): Integrated Pest Management (IPM) is a pest management approach that is both effective and environmentally sensitive.

Internode: An internode refers to the segment of a stem that is located between two nodes.

Invasive: An invasive species refers to a non-indigenous organism that poses a threat to the natural ecosystem.

Irrigation System: An irrigation system is a complex arrangement of pipes, pumps, valves, and emitters that is specifically intended to distribute water to crops in a regulated and efficient manner.

Landscaping: Landscaping refers to the deliberate and artistic process of planning and executing the design of outdoor areas. The procedure typically commences with a landscaping plan.

Layering: Layering is a method of asexual plant propagation. The nascent plant is provisionally affixed to a maternal plant during the process of root formation. Natural layering occurs when a branch comes into contact with the soil and develops new adventitious roots.

Leaching: Leaching refers to the process of nutrients being drained away from the soil. This also applies to potted plants, as they require frequent fertilization in order to promote optimal plant growth.

Leaf mold: Leaf mold is created through the decomposition of leaves, resulting in the formation of organic compost.

Leggy: Leggy refers to the elongated and thin growth of a plant, typically caused by a lack of adequate light.

Legume: A legume is a type of seed pod. Examples of legumes include beans, peas, and lentils.

Lenticel: A lenticel refers to the openings found in a woody stem that facilitate the exchange of gases between the plant and the atmosphere.

Light Feeders: Light Feeders are plants that thrive with minimal fertilizer for best performance. Excessive feeding of organisms that consume little amounts of food might lead to the accumulation of hazardous substances.

Loose Flowers: Loose flowers refer to flowers that are typically plucked without their stalks and are commonly used for adorning hair, worshiping deities, and making garlands.

Lopping: Lopping is the act of decreasing the amount of foliage in shadow trees to allow for increased sunlight penetration.

Macro-nutrient: The essential macro-nutrients for optimal plant growth are nitrogen, phosphorus, and potassium.

Manure: Manure refers to the excrement of animals. A natural soil fertilizer and conditioner.

Meristem: Meristem is a type of plant tissue in which cells undergo active division.

Metamorphosis: Metamorphosis refers to the biological process through which insects undergo a transition from their immature stage to their adult form.

Microbe: A microbe is a small organism, typically invisible to the naked eye.

Microclimate: Microclimate refers to the specific climate conditions that exist inside a smaller region or area, which is part of a larger overall climate. Especially important in a city setting, as it can impact the development and performance of plants. For instance, barrier walls can have an impact on the amount of sunlight reaching an area and can also influence the direction and flow of wind.

Micrografting: Micrografting refers to the process of grafting small plant components under sterile and carefully controlled environmental conditions. Micrografting has mostly been employed in citrus, apple, and plum cultivation with the purpose of generating plants that are free from viral infections.

Micro-nutrient: Micro-nutrients, such as boron (B), zinc (Zn), manganese (Mn), iron (Fe), copper (Cu), molybdenum (Mo), and chlorine (Cl), are vital for the body, but they are only needed in very small quantities.

Monocot: Monocotyledonous plants have a single cotyledon in their seeds. For instance, let's consider the topic of grass.

Monoecious: Monoecious refers to a plant that possesses both male and female flowers on the same individual.

Mosaic: Mosaic is a viral infection that causes leaves to have yellow, white, and green patches. As a plant virus, there is no remedy available, and the most effective approach would be to eliminate and eliminate affected plants.

Mulch: Mulch refers to a layer of material, typically organic but sometimes inorganic, that is applied over the soil to provide protection. Typically employed to confine the development of unwanted plants and aid with the preservation of moisture. Mulch is a layer of organic or inorganic material that is deposited on the soil surface surrounding plants. Its purpose is to conserve moisture, prevent the growth of weeds, and regulate the temperature of the soil.

Mycelium: Mycelium refers to a dense network of branching, thread-like hyphae that form the main body of fungus.

Mycorrhiza: Mycorrhiza refers to a type of fungi that establishes a mutually beneficial association with the roots of plants.

Naturalizing: Naturalizing refers to the process of planting bulbs that have the ability to return and bloom year after year without needing to be replanted.

Necrosis: Necrosis refers to the demise of plant tissue.

Nematode: Nematodes are little, worm-like creatures that have the potential to induce plant diseases.

Node: Location on a stem to which a leaf or branch is attached. Most pruning should be done just above a node.

Nose: The nose has a bulbous tip.

Notching: Notching refers to the act of partially removing a section of a branch above a latent lateral bud. This technique is commonly used on plants such as figs and apples.

Nursery: A nursery is a specialized facility that focuses on the propagation and cultivation of plants, nurturing them until they reach a stage when they can be successfully transplanted into various landscapes. Nurseries cultivate and generate a variety of plants, including ground-cover plants, herbaceous perennials, flowering shrubs, evergreens, deciduous shade trees, and decorative trees, with the purpose of selling them to both individual customers and businesses.

Nymph: A nymph is the juvenile stage of insects.

Ornamenatal Horticulture: Ornamental Horticulture is the field of study that focuses on numerous categories of decorative plants that are utilized for enhancing the appearance of both indoor and outdoor gardens.

Ovule: An ovule is the reproductive structure in flowering plants that develops into a seed. Parthenocarpic refers to the process of fruit production occurring without the need for pollination. Here are several examples: pineapple, banana, watermelon.

Parthenocarpic: Parthenocarpy is the process by which seedless cultivars are created. Parthenocarpic refers to the process of fruit production occurring without the need for pollination. Here are several examples: pineapple, banana, watermelon. Parthenocarpy is the process by which seedless cultivars are created.

Patch budding: Patch budding involves the full removal of a rectangular section of bark from a one-year-old seedling stock. This is then replaced with a similar section of bark that contains a bud of the desired variety. It is effectively employed in species with a dense outer layer of bark, such as aonla, bael, jamun, guava, walnut, and pecan nut.

Pathogen: A pathogen is a general term for an organism that causes illness in plants. Some examples of organisms include fungi, bacteria, and nematodes.

Peat moss: Peat moss refers to the process of creating a growth medium using sphagnum moss. Due to its low pH (acidic) level, it is a suitable amendment for plants such as blueberries and azaleas. Peat has negligible organic fertilizer value. In order to mitigate the ongoing degradation of peatland, the United Kingdom will implement a prohibition on the selling of peat moss to gardeners by the year 2024.

Pedicel: A solitary bloom borne on a stalk.

Peduncle: Stem of a flower.

Perennials: Perennial plants have a lifespan beyond two years, which is longer than biennial plants.

Perfect flower: Both male and female reproductive structures are present.

Perlite: An organic substance derived from volcanic material that enhances the process of aeration, water retention, and drainage when incorporated into soil as a supplement.

Pesticide: Pesticides are substances utilized to manage and regulate pests.

pH: pH is a quantitative measure of the acidity or alkalinity of a substance. Using a numerical scale ranging from 1 to 14, a value of 7 represents neutrality. Values below 7 indicate acidity, while values over 7 indicate alkalinity.

Phloem: Phloem is the vascular tissue responsible for transporting nutrients to the leaves and roots. Situated on the external surface of the cambium.

Photoperiodism: Photoperiodism refers to the way in which plants react to the duration of light exposure they get. Long day plants necessitate a greater amount of light exposure compared to short day plants. Day neutral plants exhibit a photoperiodic response in which they require an equal amount of exposure to both light and darkness.

Photosynthesis: Photosynthesis is the biological process in which green plants utilize sunlight to combine carbon dioxide and water, resulting in the synthesis of food.

Phototropism: Phototropism refers to the movement of plants in reaction to light.

Phytosanitary Measures: Phytosanitary measures refer to the practices and policies implemented to avoid the introduction and spread of pests and diseases in horticultural crops.

Pinching: Pinching refers to the act of removing a developing stem, typically the main bud, in order to encourage the formation of several branches and a more compact shape. Frequently employed in the cultivation of bonsai trees.

Pistil: The pistil refers to the collective female reproductive organs of a flower, which include the ovary, stigma, and style.

Pith: The pith refers to the innermost section of a tree, situated near the middle of the trunk.

Plasticulture: Plasticulture is a widely used advanced horticulture technology in modern times. Plastics are utilized in several ways in commercial horticulture production. The word 'Plasticulture' refers to the process of utilizing plastics for commercial horticultural output. Plastics are utilized in horticulture for many purposes such as Protected Cultivation (which involves the use of greenhouse buildings, high and low tunnels), Plastic Mulching, and Plastic Lining. Plasticulture enhances the economic efficacy of production systems and facilitates efficient use of water and energy. Plasticulture mitigates temperature and moisture changes, while also aiding in the management of insect and disease infestations.

Pollen: Pollen refers to tiny particles generated by the stamen of a flower, which stimulate the pistil to develop seeds.

Pollination: Pollination is the process of transferring pollen grains from the male anther to the female stigma.

Precision farming: Precision farming is a farming approach that emphasizes the use of cutting-edge technologies and advancements in crop production. This indicates that the cultivator have precise knowledge on how to effectively manage the production process in order to attain the highest possible output and quality of the specific crop. By optimizing input efficiency and maximizing output while minimizing energy waste, this approach not only enhances environmental sustainability but also boosts profitability. Precision farming encompasses all advanced horticulture tools.

Precision Irrigation: Precision Irrigation is a technology-based method that enhances the efficiency of water and nutrient distribution, minimizing wastage and maximizing crop productivity.

Propagation: Propagation refers to the process of generating new plants using a range of different techniques. It can reproduce sexually through seeds and asexually through methods such as cuttings.

Protected\Greenhouse cultivation: Greenhouse cultivation refers to a method of cultivation in which the microclimate surrounding the plant is completely or partially regulated in order to shield the plant from unfavorable weather conditions. This not only offers a greater output in confined areas but also enables the cultivation of crops in difficult conditions and during non-traditional growing seasons. The majority of cut flowers used for export are sourced from technologically advanced floricultural facilities.

Prune - Using pruning shears, scissors, a knife, or loppers to shape or rejuvenate a plant, not to increase branching. Generally pruning is much more drastic than pinching. Pruning is most commonly used on shrubs, trees, and perennials.

Raceme – A flower cluster with the separate flowers attached by short equal stems.

Radicle – First part of the seedling to emerge from the seed.

Resistance – The ability of plants to combat pathogens. Developing disease resistant plants is a prominent part of modern horticulture.

Respiration – The absorption of oxygene molecules to produce water, carbon dioxide and energy.

Rhizome – A modified stem that send out shoots and roots from its nodes. Examples, canna lillies, bearded iris, ginger.

Ring budding: In the process of ring budding, the stock is totally girdled by removing a full ring of bark. A corresponding section of bark, which includes a bud, is extracted from the bud stick and then put onto the rootstock. Both the scion and stock in this grafting process must be of equal size. It is employed in peach, plum, ber, mulberry, and other fruits.

Roguing: Removal of inferior and diseased plants.

Root hairs: Cylindrical root extensions used to absorb food and water.

Root rot: Tissue decay mostly due to excessive water.

Root: Underground part of a plant. Anchors, absorbs minerals and sometimes stores food.

Rootstock: The root system and part of a tree where another tree (the scion) is grafted on. Very common on fruit trees and roses.

Sapwood : Outer, living layers of secondary wood delivering water through the xylem.

Scaffold branching : The primary branches emerging from the trunk.

Scape: A long, leafless flower stalk emerging directly from the roots. Example, daffodil.

Scarification: Delicately weakening the seed coat to promote germination. The mechanical method is the most commonly used.

Scion: The detached shoot that is propogated onto rootstock by grafting.

Scorch: Browing and drying out of leaf tips and margins.

Scoring: Making a shallow cut around the trunk or branches of fruit trees to increase yield.

Secondary growth: Is characterized by the increase in thickness (girth) as a result of cell division in lateral meristems (cambium).

Secondary mineral nutrients: The elements present are boron, copper, iron, chlorine, manganese, zinc, and molybdenum. Plants require these substances in relatively modest quantities, and they are generally abundant and easily accessible in most soils.

Self-pollination: Pollination from the anther to the stigma takes place on the same plant.

Shade net cultivation: Shade net agriculture utilizes nets to offer shade to plants, safeguarding them from excessive sunlight and heat. This approach is especially beneficial in arid and warm areas, where elevated temperatures have the potential to harm agricultural produce.

Simple layering: When a pliable branch flexes and makes contact with the ground, roots are generated, leading to the growth of a new plant. Asexual propagation refers to the reproduction of plants without the involvement of sexual organs or processes.

Slow-Release Fertilizer: The fertilizer is in the form of pellets and is released gradually, primarily through the activity of microbes, which is influenced by soil temperatures.

Smudging: It is the act of smoking mango trees, a common technique in the Philippines to yield crops during the off-season. There is no evidence to suggest that smudging of Mango trees in India causes early flowering.

Soil conditioner: Soil additives for enhancing soil structure. For instance, gypsum.

Sphagnum moss/sphagnum peat: Soil additives for enhancing soil structure. For instance, gypsum.

Spore: Particles that allow fungi to reproduce.

Sprinkler Irrigation: A technique that sprays water over the crops in the form of droplets, mimicking natural rainfall, suitable for various crops and field sizes.

Stamen: Pollen producing reproductive organ (male) of a flower.

Stem: Bears the leaves and flowers.

Stigma: The apex of the flower pistil (female) where pollen is deposited.

Stipule: A small outgrowth found on the base of a leafstalk. Example, roses.

Stratification: Exposing seeds to moist cool conditions which aide germination.

Subsurface Irrigation: A technique that delivers water below the soil surface directly to the root zone, reducing evaporation and surface runoff.

Succulents: Have thick fleshy leaves for sustained water retention. They thrive in arid climates.

Sucker: Shoots emerging and growing from the roots of trees and shrubs.

Surface Irrigation: A method where water flows over the soil surface and is distributed across the field by gravity, commonly used for rice paddies.

Susceptible or Susceptibility: This is the antithesis of resistance. The plant lacks the capacity to withstand or mitigate the harm caused by a disease, thus resulting in severe consequences for the plant.

Systemic: Transportation of substances within the plant's vascular system. For instance, systemic herbicides go through the plant's vascular system, causing the death of the entire plant.

Terminal bud: The bud at the tip of a stem.

Thatch: Mostly dead, tightly intermingled material that develops on neglected lawns. Removing this material is called dethatching.

Thinning: A broad term to remove plant and plant material to avoid overcrowding.

Thorn: Sharp modified stems that grow from buds and are connected to the vascular system. Example, hawthorn.

Tilth: The physical condition of soil and suitability for planting and growing crops.

Time Release Fertilizer: Controlled Release Fertilizer, sometimes known as CRF, is a term used to describe this type of fertilizer. Fertilizer is available in pellet form and is an enhanced iteration of Slow-Release Fertilizer. Fertilizer is released solely depending on soil temperature, rather than microbial activity, and is generally more precise compared to Slow-Release Fertilizer.

Tissue culture: The cultivation of plants from cell level in a sterile environment. Also called micropropagation.

Tolerance: This term is used to describe situations where disease is not the underlying cause of plant stress. Phenomena such as heat stress, drought stress, and pest infestations might be categorized as instances of tolerance. Tolerance, in this context, refers to the plant's ability to withstand and endure a specific stress without suffering significant damage to its development or performance.

Tongue grafting: This method is practiced as whip grafting except one additional reverse cut is made on both scion and rootstock, so that cambial contact will be more with more success percentage.

Top working: Grafting established trees into another variety by inserting grafts into its branches.

Training: Pruning and staking plants to adapt to a particular area or shape.

Translocation: Movement of substances through the plant vascular system.

Transpiration: Transpiration refers to the process by which water is transported and evaporated from plants. The majority of water consumed, approximately 97 to 99.5 per cent, is expelled through transpiration via the stomata.

Transplanting: The process of moving young seedlings or plants from a nursery or seedbed to their final growing location.

Variegated Foliage: Foliage with different colors, usually but not always random, alternating on the foliage.

Vascular cambium: A layer of meristem cells between the primary phloem and xylem.

Vascular system: Cells that distribute nutrients and water throughout the plant.

Vector: Usually, insects that tranfer the spread of pathogens to plants.

Vegetative reproduction – The asexual reproduction of plants. Example, cuttings and grafting.

Vermiculite: A mineral that has undergone heat treated expansion and is used as a soil ammendment and rooting agent for plants.

Vertical farming: Vertical farming involves cultivating plants in multiple tiers, utilizing artificial illumination and a regulated atmosphere. This technique is especially advantageous in densely populated urban regions where there is little space, as it enables the development of plants in a concentrated manner.

Warm-season Grass: The growth of these grasses will not commence until the middle to late part of spring, or possibly even the beginning of summer. Their primary growth and blooming occur during periods of high temperatures. Typically, they will undergo a transformation and assume various hues of brown during the winter season.

Water sprout: Shoots that emerge from the main stem of a tree or its branches. This is frequently caused by extensive trimming, as these shoots have a tendency to be frail.

Watering: Plants vary in their water requirements and can be classified into five broad groups. These categories are primarily applicable to potted plants but can also be used to plants grown in the ground.

Waterlogged: Soil saturated with water that lacks oxygen.

Wet Feet: Plants that are constantly exposed to moist soil, either in a container or in the landscape, are commonly described as having "wet feet." Some plants have roots that are averse to persistent moisture, and we could describe this as the plant having an aversion to saturated roots. In contrast, several plants have roots that can tolerate being consistently saturated with water, and we can describe these plants as being tolerant of having wet roots.

Xeriscape: Xeriscaping is a landscaping technique that utilizes drought-tolerant plants to promote water conservation.

Xylem: The xylem is a type of vascular tissue found on the inner side of the cambium. Its primary function is to transfer water and dissolved nutrients in an upward direction.

References

Abegunde VO, Sibanda M, Obi A. The dynamics of climate change adaptation in Sub-Saharan Africa: A review of climatesmart agriculture among small-scale farmers. Climate. 2019; 7(11): 132.

Aditya P, Rao V, Mohapatro S, Chandra V, Nanda C, Suman S. Future trends in protected cultivation: A Review; 2023.

Alansari Z, Anuar NB, Kamsin A, Soomro S, Belgaum MR, Miraz MH, *et al.,* Challenges of internet of things and big data integration. In Emerging Technologies in Computing: First International Conference, iCETiC 2018, London, UK, August 23–24, 2018, Proceedings 1. Springer International Publishing. 2018; 47- 55

Ameta KD, Kaushik RA, Dubey RB, Rajawat KS. Protected cultivation-An Entrepreneurship for modern agriculture. Biotech Today: An International Journal of Biological Sciences. 2019; 9(1): 35-40.

Ameta KD, Kaushik RA, Dubey RB, Rajawat KS. Protected cultivation-An Entrepreneurship for modern agriculture. Biotech Today: An International Journal of Biological Sciences. 2019; 9(1): 35-40.

Anonymous (1998). The Mango. Tech. Bull., CISH, Lucknow.

Anonymous (2001): Hi-tech Horticulture Policy paper.

Anonymous (2009): ICAR Data book.

Anonymous. Post-harvest Management of Fruit and Vegetables in the Asia-Pacific Region. Food and Agriculture Organization of the United Nations Agricultural and Food Engineering Technologies Service, Viale delle Terme di Caracalla, 00100 Rome, Italy, 2006. ©APO 2006, ISBN: 92-833-7051-1

Anung K. hal and Thomas. F. Schere: (2003): Introduction to Micro-irrigatiion.

Asati BS, Yadav DS. Diversity of horticultural crops in north eastern region. ENVIS Bulletin: Himalayan Ecology. 2004; 12(2): 1-10.

Atzberger, C. (2013). Advances in remote sensing of agriculture: Context description, existing operational monitoring systems and major information needs. Remote sensing, 5(2), 949-981.

Baeza EJ, Stanghellini C, Castilla N. Protected cultivation in Europe. In International Symposium on High Tunnel Horticultural Crop Production. 2011, October; 987: 11-27.

Barritt BH, Konishi BS, Dilley MA. Tree size, yield and biennial bearing relationships with 40 apple rootstocks and three scion cultivars. Acta Hort 1997; 451: 105-112.

Baylis A. Advances in precision farming technologies for crop protection. Outlooks on Pest Management. 2017; 28(4): 158- 161.

Benke K, Tomkins B. Future foodproduction systems: Vertical farming and controlled-environment agriculture. Sustainability: Science, Practice and Policy. 2017; 13(1): 13-26.

Boswell VR. Pepper production (No. 276). US Department of Agriculture, Agricultural Research Service; 1964.

Bradley, S. 2005. The pruner's bible: A step-by-step guide to pruning every plant in your garden. Rodale,

Bruchou C, Génard M. A space–time model of carbon translocation along a shoot bearing fruits. Annals of Botany 1999; 84: 565-576.

Burondkar, M.M., Gunjate, R.T., Nagdum, M.B. and Govekar, M.A. (2000). Rejuvenation of old and overcrowded Alphonso mango orchard with pruning and use of paclobutrazol. Acta Hort. 509: 681-86.

Castilla N. Current situation and future prospects of protected crops in the Mediterranean region. In International Symposium on Mediterranean Horticulture: Issues and Prospects. 2000, October; 582: 135-147.

Chadha, K. L. (2001). Handbook of horticulture. ICAR New Delhi.

Chan BG, Cain JC. The effect of seed formation on subsequent flowering in apple. Proc. Am. Soc. Hort. Sci 1967; 91: 63-68.

Chattopadhyay, T.K. (1996). A Text Book on Pomology. Vol. II. Kalyani Publishers, New Delhi.

Chaturvedi BK, Raj LCA. Agricultural Storage Infrastructure in India: An Overview. IOSR Journal of Business and Management, 2015; 17(5): 37-43.

Chris B Watkins, Jacqueline F Nock. Production Guide for Storage of Organic Fruits and Vegetables, Department of Horticulture, Cornell University, NYS IPM Publication No. 2012, 10.

Communications in Biometry and Crop Science, 11, 31-50.

Corelli Grappadelli L, Lakso AN, Flore JA. Early season patterns of carbohydrate partitioning in exposed and shaded apple branches. J Amer. Soc. Hort. Sci 1994; 119: 596-603.

Corelli Grappadelli L, Ravaglia G, Asirelli A. Shoot type and light exposure influence carbon partitioning in Peach cv. 'Elegant Lady'. J Hort. Sci 1996; 71: 533-543.

Criley, R.A. 1997. Bougainvillea. In: M.L. Gaston, S.A. Carver, C.A. Irwin, and R.A. Larson (eds.), Tips on

Dakshinamurti, C., Krishnamurthy, B., Summanwar, A. S., Shanta, P. and Pisharoty, P. R. (1971). Remote sensing for coconut wilt. Proceedings of International Symposium on Remote Sensing of Environment, vol (1), Ann Arbor, MI: 25-29.

Dalai S, Tripathy B, Mohanta S, Sahu B, Palai JB. Green-houses: Types and Structural Components. Protected Cultivation and Smart Agriculture. New Delhi Publishers. New Delhi, India; 2020.

D'antonio CARLA, Meyerson LA. Exotic plant species as problems and solutions in ecological restoration: A synthesis. Restoration Ecology. 2002; 10(4): 703-713.

De Candolle A. Origin of cultivated plants. Cambridge University Press 1982.

De Gelder A, Dieleman JA, Bot GPA, Marcelis LFM. An overview of climate and crop yield in closed greenhouses. The Journal of Horticultural Science and Biotechnology. 2012; 87(3): 193-202.

De LC, Singh DR. Floriculture industries, opportunities and challenges in Indian hills. International Journal of Horticulture. 2016; 6(13): 1-9.

De Pascale S, Maggio A. Sustainable protected cultivation at a Mediterranean climate. Perspectives and challenges. In International Conference on Sustainable Greenhouse Systems-Greensys. 2004, September; 2004(691): 29-42.

DeJong TM. Peach Tree Vigor Is a Function of Rootstock Xylem Anatomy and Hydraulic Conductance. Acta Hort 2012; 932: 483-489.

Dykun VP, Schevchuk AA. Influence of nutritional area on the productivity of bee pollinating and Parthenocarpic cucumber hybrids in the spring plastic hot houses. Ways of intensifying the vegetable growing. Kiev, Russia Fed 1990, 65-68.

Erkan M and Dogan A. (2019). Harvesting of horticultural commodities. *In Postharvest Technology of Perishable Horticultural Commodities* 129-159.

Fernández JA, Orsini F, Baeza E, Oztekin GB, Muñoz P, Contreras J, Montero JI. Current trends in protected cultivation in Mediterranean climates. Eur. J. Hortic. Sci. 2018; 83(5): 294-305.

Fernández JA, Orsini F, Baeza E, Oztekin GB, Muñoz P, Contreras J, *et al.,* Current trends in protected cultivation in Mediterranean climates. Eur. J. Hortic. Sci. 2018; 83(5): 294-305.

Fernández JA, Orsini F, Baeza E, Oztekin GB, Muñoz P, Contreras J, *et al.,* Current trends in protected cultivation in Mediterranean climates. Eur. J. Hortic. Sci. 2018; 83(5): 294-305.

Food and Agriculture Organization of the United Nations (FAO). FAOSTAT. Accessed on 15 October; 2023. Available: https: //www.fao.org/faostat/

Forshey CG, Elfving DC, Stebbins RL. Training and pruning of apple and pear trees. American Society for Horticultural Science, Alexandria, Virginia, USA 1992.

Forshey CG, Elfving DC. The relationship between vegetative growth and fruiting in apple trees. Hort. Rev 1989; 11: 229-287.

Gary C, Jones JW, Tchamitchian M. Crop modelling in horticulture: state of the art. Scientia Horticulturae. 1998; 74(1-2): 3-20.

Gilman, E.F. 1999. Bougainvillea spp. Cooperative Extension Service, University of Florida, Fact Sheet

Gindling TH, Newhouse D. Selfemployment in the developing world. World Development. 2014; 56: 313-331.

Gomiero T, Paoletti MG, Pimentel D. Energy and environmental issues in organic and conventional agriculture. Critical Reviews in Plant Sciences. 2008; 27(4): 239-254.

Gorjian S, Calise F, Kant K, Ahamed MS, Copertaro B, Najafi G, *et al.,* A review on opportunities for implementation of solar energy technologies in agricultural greenhouses. Journal of Cleaner Production. 2021; 285: 124807.

Granatstein D, Kirby E, Willer H. Current world status of organic temperate fruits. In Organic Fruit Conference. 2008, June; 873: 19-36.

Gruda N, Tanny J. Protected crops. Horticulture: Plants for People and Places, Production Horticulture. 2014; 1: 327-405.

Guo FC, Fujime Y, Hirose T, Kato T. Effect of the number of training shoots, raising period of seedlings and planting density in Dhillon *et al.,* JEAI 2017; 18(1): 1-5. Article no. JEAI.36363 5 growth, fruiting and yield of sweet pepper. J Jpn Soc Hortic Sci 1991; 59: 763-770.

Hackett, W.P., R.M. Sachs, and J. DeBie. 1972. Growing bougainvillea as a flowering pot plant. California

Hamilton, D.F., and J.T. Midcap. 2003. Propagation of woody ornamentals by cuttings. Cooperative Extension Service, University of Florida.

Hanan JJ. Greenhouses: Advanced technology for protected horticulture. CRC Press; 2017.

Hartman, H. T., D.E. Kester., F.T. Davies and R.L. Geneve (1997): Plant Propagation-Principles and Practices.

Hasan M. Protected Cultivation and Drip fertigation technology for sustainable food production. International Journal of Economic Plants. 2016; 3(3): 102-106.

Hawai'i at Mänoa, College of Tropical Agriculture and Human Resources (CTAHR).

Hebert M. Greenhouse cucumber production 1998. Available: http: //www.uaf.edu

https: //aggiehorticulture.tamu.edu/extension/newsletter/vpmnews/mar01/ art2mar.html

https: //agritech.tnau.ac.in/govt_schemes_services/govt_serv_schems_nadp_tnau per cent 20_precisionfarming.html

https: //grindgis.com/remote-sensing/

https: //www.agrifarming.in/horticulture-precision-farming-technology-advantages

Huang Y, Li W, Zhao L, Shen T, Sun J, Chen H, *et al.,* Melon fruit sugar and amino acid contents are affected by fruit setting method under protected cultivation. Scientia Horticulturae. 2017; 214: 288-294.

Imler CS. Quantifying temperature effects in controlled environment agriculture leafy greens and culinary herbs (Doctoral dissertation, Iowa State University); 2020.

Iredell, J. 1994. Growing bougainvilleas. Simon and Schuster Australia. East Roseville, NSW, Australia.

Jewett T, Jarvis W. Management of the greenhouse microclimate in relation to disease control: A review. Agronomie. 2001; 21(4): 351-366.

Kanwar MS. 5 High-Altitude Protected Vegetable Cultivation–A Way for Sustainable Agriculture. Applied Agricultural Practices for Mitigating Climate Change. 2019; 2: 51.

Kaushal S, Singh V. Potentials and prospects of protected cultivation under Hilly Conditions. Journal of Pharmacognosy and Phytochemistry. 2019; 8(1): 1433-1438.

Khandetod YP. (2019). Mechanization in Horticulture Crops: Present Status and Future Scope. *Advanced Agricultural Research and Technology Journal* **3**: 92-103.

Knowler D, Bradshaw B. Farmers' adoption of conservation agriculture: A review and synthesis of recent research. Food Policy. 2007; 32(1): 25-48.

Kumar D, Singh B. Vegetables cultivation under protected conditions. Progressive Agriculture. 2020; 20(1and2): 148-152.

Kumar D, Singh B. Vegetables cultivation under protected conditions. Progressive Agriculture. 2020; 20(1and2): 148-152.

Kumar S, Meena RS, Jakhar SR, Jangir CK, Gupta A, Meena BL. Adaptation strategies for enhancing agricultural and environmental sustainability under current climate. Sustainable agriculture. Scientific Publisher, Jodhpur. 2019; 226-274.

Lal, B., Rajput, M.S., Rajan, S. and Rathore, D. S. (2000). Effect of rejuvenation of old mango trees, Indian J. Hort. 57: 240-42.

Lamichhane P, Adhikari J, Poudel A. Protected cultivation of horticultural crops in Nepal: Current practices and future needs. Archives of Agriculture and Environmental Science. 2023; 8(2): 268- 273.

Marra M, Pannell DJ, Ghadim AA. The economics of risk, uncertainty and learning in the adoption of new agricultural technologies: where are we on the learning curve?. Agricultural Systems. 2003; 75(2- 3): 215-234.

Martínez-Gómez P, Rahimi Devin S, Salazar JA, López-Alcolea J, Rubio M, Martínez-García PJ. Principles and prospects of prunus cultivation in greenhouse. Agronomy. 2021; 11(3): 474.

More TA, Chandra P, Singh JK. Cultivation of cucumber

Muthukumar, P. and Selvakumar, R. (2013). Glaustus Horticulture. New Vishal Publication, New Delhi.

Nageswara Rao, P. P., Ravishankar, H. M. and Uday Raj, K. N. (2004). Production estimation of horticultural crops using IRS–1d liss–iii data. J. Indian Soc. Remote Sens., 32: 393-398.

Nath A, Bagchi B, LK Misra, Bidyut C Deka. Changes in post-harvest phytochemical qualities of broccoli florets during ambient and refrigerated storage. Food Chemistry, 2011; 127: 1510-1514.

Nath A, Bagchi B, VK Verma, H Rymbai, AK Jha, BC Deka. Extension of Shelf Life of Tomato Using KMnO4 as Ethylene Absorbent. Indian Journal of Hill Farming. 2015; 28(1): 77-80.

Nath A, Deka BC, D Paul, LK Mishra. Ambient storage of capsicum under different packaging materials. Bioinfolet, 2010; 7(3): 266-270.

Nath A, Deka Bidyut C, Jha AK, Paul D, Misra LK. Effect of slice thickness and blanching time on different quality attributes of instant ginger candy. Journal of Food Science and Technology, 2013; 50(1): 197-202.

Nath A, Deka Bidyut C, Ngachan SV. A process for producing tuity-fruity from Chow chow. The Patent Office Journal, 12/12/2014, 2014, 15201.

Nath A, K Barman, S Chandra, P Baiswar. Effect of plant extracts on quality of Khasi mandarin (Citrus reticulata Blanco) fruits during ambient storage. Food and Bioprocess Technology, 2013; 6(2): 470-474.

Nath A, Mangaraj S, Goswami TK, Chauhan J. Post-harvest management and production of important horticultural crops. Scientific Publishers (India), Jodhpur, India. ISBN: 978-81-7233-948-7. 2016, 1-436.

Nath A. Post-harvest Management and Value Addition in Horticultural Crops: Scope for Entrepreneurship Development with Special Reference to North-East India. Published in Horticulture for economic prosperity and nutritional security in

21st century. Edited by T. K. Hazarika and B.P. Nautiyal Published by Westville Publishing House, New Delhi. 2013, 207-219.

Navarro S, Noyes R. The Mechanics and Physics of Modern Grain Management. CRC Press, Boca Raton, London, New York, Washington DC. 2002, 647.

Navarro, L and J Juarez (2007): Shoot-tip grafting in vitro: Impact in the citrus industry and Research application. Citrus Genetic, Breeding and Biotechnology.353-365.

Navarro,L. and T. Juorez (1975): Shoot tip grafting in Vitro: Impact in citrus industry and Research Application

Neal, M.C. 1965. In gardens of Hawaii. Bishop Museum Press, Honolulu.

Negi, S.S. (2000). Mango production in India. Acta Hort. 509: 69-78.

Negi, S.S., Rajan, S. and Kumar, Ram. (2000). Developing mango varieties through hybridization. Acta Hort. 509: 159-60.

Nordey T, Basset-Mens C, De Bon H, Martin T, Déletré E, Simon S, *et al.*, Protected cultivation of vegetable crops in sub-Saharan Africa: limits and prospects for smallholders. A review. Agronomy for Sustainable Development. 2017; 37: 1-20.

Ottman J. The new rules of green marketing: Strategies, tools, and inspiration for sustainable branding. Routledge; 2017.

Pachiyappan P, Kumar P, Reddy KV, Kumar KNR, Konduru S, Paramesh V, *et al.*, Protected cultivation of horticultural crops as a livelihood opportunity in western India: An economic assessment. Sustainability. 2022; 14(12): 7430.

Pal A, Adhikary R, Shankar T, Sahu AK, Maitra S. Cultivation of cucumber in greenhouse. Protected cultivation and smart agriculture. New Delhi Publishers, New Delhi, India; 2020. DOI: 10, 139-145

Pallas Jr JE, Michel BE, Harris DG. Photosynthesis, transpiration, leaf temperature, and stomatal activity of cotton plants under varying water potentials. Plant Physiology. 1967; 42(1): 76-88.

Pancharatnam P. Agriculture Letters. Agriculture Letters. 4.

Paradiso R, Proietti S. Light-quality manipulation to control plant growth and photomorphogenesis in greenhouse horticulture: The state of the art and the opportunities of modern LED systems. Journal of Plant Growth Regulation. 2022; 41(2): 742-780.

Peet MM, Welles G. Greenhouse tomato production. In Tomatoes. Wallingford UK: CABI Publishing. 2005; 257-304.

Pertuit, A. 1999. Bougainvilleas. Cooperative Extension Service, Clemson University, Home and Garden Information Center HGIC 1553.

Peter, K.V. (2007). Fruit Crops. New India Publishing Agency, New Delhi.

Peterson DL. (2005). Harvest mechanization progress and prospects for fresh market quality deciduous tree fruits. *HortTechnology* **15**(1): 72-75.

Prakash P, Kumar P, Kar A, Kishore P, Singh AK, Immanuel S. Protected cultivation in Maharashtra: Determinants of adoption, constraints, and impact. Agricultural Economics Research Review. 2021; 34(2).

Prasad.S., and U.Kumar(2005): Greenhouse Management for Horticultural crops.

Ragunathan, V., Pawar, A.D., Misra, M.P. and Singh, J. (2002). Integrated management of pests in horticultural crops. In: Approaches for Sustainable Development of Horticulture.

Rahman F, Abid K, Schmidt C, Pfaff G, Koenig F. Interference pigment coated solar cells for use in high radiant flux environments. JJMIE. 2010; 4(1).

Rahman MM, Khan I, Field DL, Techato K, Alameh K. Powering agriculture: Present status, future potential, and challenges of renewable energy applications. Renewable Energy. 2022; 188: 731-749.

Raina RS, Joseph K, Haribabu E, Kumar R. Agricultural innovation systems and the coevolution of exclusion in India. Systems of Innovation for Inclusive Development Project; 2009.

Ram, S. and Tripathi, P.C. (1993). Effect of cultar on flowering and fruiting in high density Dashehari mango trees. Indian. J. Hort. 50: 292-95.

Ram, S., Bist, L.D. and Sirohi, S.C. (1989). Internal fruit necrosis in mango and its control. Acta Hort. 231: 305-13.

Ram, S., Singh, C.P. and Kumar, S. (1997). A success story of high-density orcharding in mango. Acta Hort. 455 (1): 375-82.

Rasheed R, Ashraf MA, Iqbal M, Hussain I, Akbar A, Farooq U, Shad MI. Major constraints for global rice production: Changing climate, abiotic and biotic stresses. Rice research for quality improvement: Genomics and genetic engineering: Breeding Techniques and abiotic Stress Tolerance. 2020; 1: 15-45.

Rashid MM. Sabgi Biggan. Rashid Publishing House, Dhaka 1999, 303.

Reddy, P. P. (2016). Sustainable crop protection under protected cultivation. Singapore: Springer.

Sabir N, Singh B. Protected cultivation of vegetables in global arena: A review. Indian Journal of Agricultural Sciences. 2013; 83(2): 123-135.

Sabir, N., and Singh, B. (2013). Protected cultivation of vegetables in global arena: A review. Indian Journal of Agricultural Sciences, 83(2), 123-35.

Santi A, Scarmuzza W, Soares D, Scarmuzza JF, Dallacort R, Krause W *et al.*, Performance and growth of conduction of Japanese cucumber in protected environment. Hortic Bras 2013; 31: 649-653.

Saxena, M., Rathore, R.P.S., Gupta, R.P., Bhargav, H. and Thakur, B. (2017). Horticultural Statistics at a Glance. Horticulture Statistics Division Department of Agriculture, Cooperation and Farmers Welfare, Ministry of Agriculture and Farmers Welfare, Government of India (9), 39.

Schoellhorn, R., and E. Alvarez. 2002. Warm climate production guidelines for bougainvillea. Cooperative Extension Service, University of Florida, ENH 874.

Selvakumar R, Glaustas Olericulture. New Vishal Publications West Patel Nagar, New Delhi, 2014, 944-945.

Sengar RS, Rani V. Opportunities and prospective of integrated development of horticulture: A review. Annals of Horticulture. 2020; 13(1): 1-8.

Sethi VP, Sharma SK. Survey of cooling technologies for worldwide agricultural greenhouse applications. Solar Energy. 2007; 81(12): 1447-1459.

Shiferaw B, Smale M, Braun HJ, Duveiller E, Reynolds M, Muricho G. Crops that feed the world Ten Past successes and future challenges to the role played by wheat in global food security. Food Security. 2013; 5: 291-317.

Shil S. Weather parameters and it's impact on agricultural production-A review. Innovative Farming. 2018; 3(4): 141-149.

Singh AK, Sabir N. Role of protected cultivation as green technology for sustainable environment: Future, Prospect, Potential, Production, Protection, and Profit. In Handbook of Research on Green Technologies for Sustainable Management of Agricultural Resources. IGI Global. 2022; 81-107.

Singh B. Advances in Protected Cultivation. New India Publishing Agency; 2014.

Singh HP, Dhankhar SS, Dahiya KK. Horticultural crops. Stadium Press (India) Pvt Limited; 2009.

Singh KK. Chapter-1 Protected Cultivation of Horticultural Crops. Chief Editor Dr. RK Naresh, 1.

Singh PK. A decentralized and holistic approach for grain management in India Current Science, 2010; 99(9): 1179-1180.

Singh, B. (2013). Protected cultivation of Vegetable crops, Kalayani publishers.

Singh, B., and Sirohi, N. P. S. (2004, June). Protected cultivation of vegetables in India: problems and future prospects. In International Symposium on Greenhouses, Environmental Controls and In-house Mechanization for Crop Production in the Tropics 710 (pp. 339-342).

Singh, H.P. and Gorakh Singh., J.C. Samuel and R.K. Pathak. (2003): Precision farming in Horticulture.

Singh, H.P., Negi, J.P., Samuel, J.C. (Eds). DAC, MOA, pp. 97-112.

Singh, J. (2011). Basic Horticulture. Kalyani Publishers, New Delhi.

Singh, K.P., Bahadur, A. (2016). Olericulture Vegetable Production and Improvement.

Singh, P.K. (2004): Hi-tech Horticulture.

Singh, R. (1969). Fruits. National Book Trust, India.

Singh, R.N., Singh, Gorakh, Rao, O.P. and Mishra, J.S. (1984). Improvement of Banarasi Langra through clonal selection. Prog. Hort. 17: 273-77

Singh, R.S. (2000). Diseases of Fruit Crops. CBS Publishers and Distributors Pvt. Ltd, New Delhi.

Singh, S.P. (1995). Commercial fruits. Kalyani Publishers, New Delhi.

Slathia D, Nisa MU, Reshi M, Dolkar T, Hussain S. Protected cultivation of ornamentals. Global Journal of BioScience and Biotechnology. 2018; 7(2): 302-311.

Srinivas, K. (2001). Microirrigation and feritigation in fruit crops. In: Microirrigation, 252-55. Singh, H.P., Kaushik, S.P., Kumar, Ashwani, Murthy, T.S., Samuel, J.C. (Eds).Central Board of Irrigation and Power.

Stassen, P.J.C., Grove, H.G. and Davie, S.J. (1999). Tree shaping strategies for higher density mango orchards. Journal of Applied Horticulture 1: 1-4.

Subin MC, Chowdhury S, Karthikeyan R. A review of upgradation of energy-efficient sustainable commercial greenhouses in Middle East climatic conditions. Open Agriculture. 2021; 6(1): 308-328.

Suthar MR, Arora SK, Bhatia AK, Dudi BS. Effect of pruning and etheral treatments on cucumber production in polyhouse. Haryana J Hortic Sci 2006; 35: 299-302.

Suthar, Ram M. Effect of pruning and ethrel application on vegetative growth and fruit yield of cucumber under greenhouse condition. Haryana J Hortic Sci 2005; 35: 92-95.

TÃ¼zel Y, Kacira M. Recent developments in protected cultivation. In VIII SouthEastern Europe Symposium on Vegetables and Potatoes. 2021, September; 1320: 1- 14.

Tack J, Yu J. Risk management in agricultural production. In Handbook of Agricultural Economics. Elsevier. 2021; 5: 4135-4231.

Tavares, J., D. Hensley, J. Deputy, D. Tsuda, and A. Hara. 1998. Bougainvillea looper. CTAHR publication IP-2.

Teho, F. 1971. Plants of Hawaii: How to grow them. The Petroglyph Press, Hilo, HI.

Tuzel Y, Leonardi C. Protected cultivation in Mediterranean region: trends and needs. J Ege Univ Fac Agric. 2009; 46(3): 215-23.

Ummyiah HM, Wani KP, Khan SH, Magray MM. Protected cultivation of vegetable crops under temperate conditions. Journal of Pharmacognosy and Phytochemistry. 2017; 6(5): 1629-1634.

Usha, K. and Singh, B. (2013). Potential applications of remote sensing in horticulture- A review. Scientia Horticulturae, 153: 71-83.

Utobo EB, Ekwu LG, Ogah EO, Nwokwu GN. Growth and yield of cucumber varieties as influenced by pruning. Continental J Agro 2010; 4: 23-27.

Van Veenhuizen R, Danso G. Profitability and sustainability of urban and periurban agriculture. Food and Agriculture Org. 2007; 19.

Vázquez PP, Ferrer C, Bueno MM, Fernández-Alba AR. Pesticide residues in spices and herbs: Sample preparation methods and determination by chromatographic techniques. TrAC Trends in Analytical Chemistry. 2019; 115: 13-22.

Watson, D.P., and R.A. Criley. 1973. Bougainvilleas. CTAHR Circular 469.

Wójtowicz, M., Wójtowicz, A., and Piekarczyk, J. (2016). Application of remote sensing methods in agriculture.

Yadav, I.S. and Rajan, S. (1993). Genetic Resources of Mangifera. In: Advances in Horticulture, vol. 1, part 1, pp. 77-93. Chadha, K.L. and Pareek, O.P. (Eds). Malhotra Publishing House, New Delhi.

Index